The Role of Microorganisms
in a
Sustainable Agriculture

The Role of Microorganisms in a Sustainable Agriculture

Selected Papers from the Second International Conference on Biological Agriculture, University of London, Wye College, Wye, Kent, UK

Edited by
J M Lopez-Real and **R D Hodges**
University of London, Wye College

A B ACADEMIC PUBLISHERS

Published in the UK by
A B Academic Publishers
P O Box 97
Berkhamstead
Hertfordshire HP4 2PX

British Library Cataloguing in Publication Data:

International Conference on Biological
   Agriculture (*2nd: 1984: Wye College*)
   The role of microorganisms in a sustainable
   agriculture: selected papers from the second
   International Conference on Biological
   Agriculture, University of London, Wye
   College, Wye, Kent, UK.
   1. Agricultural microbiology
   I. Title II. Lopez-Real, J.M. III. Hodges,
   R.D.
   630'.2'76   QR51

ISBN 0-907360-10-6

© 1986 A B Academic Publishers

Printed in Great Britain

Preface

Microorganisms in their capacities as decomposers, nutrient recyclers, symbionts or pathogens, play a crucial role in all agricultural systems. They are, however, of special relevance to the biological/organic approach where the natural processes they effect and affect are not short-circuited by the use of agrochemical inputs. The editors of the journal *Biological Agriculture & Horticulture* together with the International Institute of Biological Husbandry (I.I.B.H.) therefore agreed to mount a conference to examine the roles of microorganisms in such agricultural systems. The conference, entitled "The Role of Microorganisms in a Sustainable Agriculture", and the second in a continuing series of conferences sponsored by I.I.B.H. under the general theme of International Conferences on Biological Agriculture, was subsequently held at Wye College, the school of agriculture and horticulture of the University of London, from September 3rd. to 7th., 1984. It was attended by more that one hundred delegates from thirty countries.

The aims of the conference were two-fold. On the one hand to enable practitioners of biological farming to hear and read about the important role that microorganisms play in agro-ecosystems; and on the other hand, and possibly more importantly, for microbiologists and related scientists to consider microbial activity not just in the context of 'agriculture' but in terms of a *sustainable* agriculture—a system of food production which aspires to resource efficiency and enhanced environmental quality. Accordingly eight keynote speakers were invited to present papers on those areas of microbiology that are so frequently quoted and emphasised by enthusiasts for biological, sustainable agriculture. The papers presented cover a diverse range of topics—from soil structure and organic waste recycling to nitrogen inputs and biological control—a clear enough indication to the general reader of the crucial and pivotal role that microorganisms play in such an approach to agriculture. In certain areas much is known of the microorganisms' role and some manipulation of their activity is possible; in others it is obvious that our understanding and knowledge of basic microbially-mediated processes in the soil and at the soil/plant interface is woefully lacking; a neglected and chronically underfunded research area. The keynote speakers were specifically invited as leading, internationally-recognised experts in their respective fields of research and not necessarily for their knowledge of biological/ sustainable farming systems. For at least some of them, the question of the microbiology of a sustainable agriculture was being considered for the first time.

The keynote papers and a selection of the offered papers were published as a special double issue of *Biological Agriculture & Horticulture*. The editors would like to thank the publishers for their agreement to publish this double issue as a separate book to enable a wider audience, in both developed and developing countries, to read about, debate and hopefully participate in this exciting and formidable future challenge. We indeed hope that this volume will be the first stage in a continuing and developing assessment of the role of microorganisms, and microbiologists, in furthering the concept and practice of a sustainable agriculture.

J M Lopez-Real and
R D Hodges

Contents

Sustainable Agriculture: The Microbial Potential—The Microbiologist's Challenge

J.M. Lopez-Real

Department of Biological Sciences, Wye College (University of London), Ashford, Kent, TN25 5AH, U.K.

The nature of this introductory paper may perhaps seem unusual since it will not deal in any detail with microorganisms as such. That will be the purpose of the subsequent papers. Instead, I have decided to use this introduction to place in context the aims and purposes of the conference as originally envisaged. In other words the setting for the conference and the reasoning behind the deliberately commissioned keynote papers.

The conference title is "The Role of Microorganisms in a Sustainable Agriculture" and it is clear from a casual glance at the contents that some papers deal directly with the topic, placing an emphasis on the term "sustainable" while others are in the more general field of microorganisms and agriculture. Of course microorganisms, through their ubiquity, are clearly important for all forms of agriculture or natural ecosystems. Indeed, it became clear early on that there appeared to be a much felt need from the microbiological community for a conference that dealt with the topic of microbes and agriculture without specifying the need for a specific type of agricultural system. Perhaps that need reflects, in general, a frustrating lack of an applied agricultural microbiology that is actually utilised and manipulated within our present forms of conventional agriculture. Microbial processes operate and are vital but are not manipulated or controlled to any great extent—indeed our understanding of many of them is still extremely limited.

Participants in the conference from a microbiological or a sustainable background may quite reasonably have asked of the organising committee the following questions:

The microbiologist—Why sustainability?
The biological agriculturalist—Why microorganisms?

It is the task of this introduction to clearly suggest the relevance of a sustainable agriculture and the importance that microorganisms have to play in such a system. The role of the plant/soil microbiologist becomes or should become one of the central and pivotal operators in the system. Because of this the concept of a truly sustainable agricultural system represents an enormous challenge to the agricultural microbiologist and one that becomes daily more relevant and important for both developed and developing countries alike.

To tackle the first question requires some working definition for "sustainability" or "sustainable agriculture" as against the norm of "conventional agriculture". What we describe as conventional agriculture is, of course, of extremely recent origin. It is really only post-World War II that the advanced countries developed an agriculture heavily dependent on the use of fossil fuel-based inputs—inorganic fertilizers, pesticides, herbicides and labour saving but energy intensive farm machinery.

Undoubtedly the application of such technologies has greatly increased production and labour efficiency but there has been increasing concern throughout the world about such heavy dependence upon fossil fuel energy and the adverse effects on soil productivity and environmental quality from increased soil erosion and excessive use of agrochemical inputs. It is also worth noting that there are considerable social factors arising from the use of such intensive systems which are also generating a wide community response and restraint—the whole debate of "agriculture and the environment" which is beginning to recognise that the farmer has a greater responsibility as a keeper of the land than merely agribusiness considerations.

When we refer to "sustainable agriculture" the term is intended to encompass all the more familiar and less familiar terms that different groups have espoused over the years. Table 1 mentions a few of them. Arriving at a single definition for this is difficult as there are as many definitions as there are variants (Merrill, 1983). The USDA report (USDA, 1980) on organic farming aimed at encompassing the entire spectrum with the following definition:

> "A production system which avoids or largely excludes the use of synthetically compounded fertilizers, pesticides, growth regulators and livestock feed additives. To the maximum extent feasible organic farming systems rely upon crop rotations, crop residues, animal manures, legumes, green manures, off-farm organic wastes, mechanical cultivation, mineral-bearing rocks, and aspects of biological pest control to maintain soil productivity and tilth, to supply plant nutrients and to control insects, weeds and other pests."

Obviously such systems were the main agricultural methods before the advent of synthetic fertilizers and pesticides. However, our modern concept of biological agriculture should not be seen as a return to the past—which it is often considered to be. We are not advocating bullocks for tractors or armies of hoers for weed control. Unfortunately, biological agriculture may be

TABLE 1

Some of the descriptive names which are, or have
been, used to refer to different variations on the
theme of sustainable agriculture.

Fertility farming	Ecological farming
Organic farming	Resource efficient farming
Humus farming	Biosustainable farming
Natural farming	Biological farming
Biodynamic farming	Low energy farming
Holistic farming	Regenerative farming
Alternative farming	

incorrectly perceived in that way as it draws on many methods of traditional agriculture that have been abandoned for more than a generation. They were, in fact, abandoned not because they did not work but because they were out-competed by energy intensive inputs and agrochemical technology.

A biological agriculture aspires to being a sophisticated systems approach to agriculture that emphasises optimization rather than maximization of single crops or system components. It places primary emphasis on managing biotic interactions in ways that maximize, not yields, but agro-ecosystem stability, and minimize demands for human and industrial inputs (Friend, 1983).

Biological, sustainable agriculture should therefore represent an integration of traditional techniques in the framework of modern scientific theory together with modern advances that are appropriate. Its progress will inevitably depend on the development of new technology and a greater understanding of the fundamental processes underlying soil fertility. Biological agriculture is an unique farming system and its future will depend on developing greater understanding and development of new technology. High input conventional farming, by-passes soil biological processes through its uses of fertilizers, pesticides and mechanised preparation of soil. As a result research that is of importance with respect to a sustainable agriculture has been seriously neglected. However, in recent years the conflict between intensive, high input agriculture and the environment has resulted in the generation of a number of important conferences, initiatives and publications (Table 2).

We may now consider the question that might be put by the practitioner of a biological system—why microorganisms? It is not difficult for those of us who are agricultural microbiologists, soil scientists and plant microbiologists to see immediately the relevance of the plant and soil microflora in the shaping of such systems. Neither, in contrast to the conventional farmer, has it escaped the notice of the interested and committed practitioner and advocate of biological agriculture. Throughout their own literature they have constantly

TABLE 2

Some important conferences, initiatives and publications associated
with the recent development of biological agriculture.

1. Dutch Government report on alternative methods of agriculture. Originally published as:
    Alternatieve Lanbouw Methoden, Centre for Agricultural Publication and Documentation,
    Wageningen, 1976. English translation in Boeringa (1980).

2. International Federation of Organic Agriculture Movements (IFOAM) conferences:
        Switzerland, 1977. (Besson & Vogtmann, 1978).
        Montreal, 1978. (Hill & Ott, 1982).
        Brussels, 1980. (Hill & Ott, 1982).
        Boston, 1982. (Lockeretz, 1983).
        Witzenhausen (W. Germany), 1984. (In press).

3. U.S. Department of Agriculture Report on Organic Farming (USDA, 1980).

4. I.I.B.H. 1st. International Conference on Biological Agriculture, 1980. (Stonehouse, 1981).
5. American Society of Agronomy Symposium on Organic Farming, 1981. (A.S.A., 1984).
6. Biological Agriculture & Horticulture (international scientific journal) first published 1982.
7. Resource-Efficient Farming Methods for Tanzania. Proceedings of a Workshop, May 1983.
    Rodale Press.
8. Soil Biological Processes and Tropical Soil Fertility. A Proposal for a Collaborative
    Programme of Research. Biology International, Special Issue (1984).
9. Michigan State University Conference on Sustainable Agriculture, 1984. (Edens *et al.*, 1985).

Other initiatives that have taken place recently include the creation of a Chair of Alternative
Agriculture at Kassel University, West Germany (Prof. H. Vogtmann) and a Department of
Alternative Agriculture at Wageningen University, The Netherlands (Prof. J.D. van Mansvelt).

proclaimed the major role of microorganisms and the soil microbiota—and
perceive of them not only as agents of disease but as the principal agents of soil
health. Indeed a basic philosophy to such practitioners, and of importance to
subsequent development of management strategies and principles, has been
the concept of "feeding the soil" rather than the plant. This implies movement
of nutrients as quickly as possible from crop residues, manures or composts
into the soil biomass and only then by mineralization into soil solution and
plant uptake. Our future quest is the potential that they may afford by
manipulation to optimize the system. Another valuable service that the
conference might achieve is to place the microbial role and its potential into a
proper perspective for the layman—so that claims based on questionable
formulations and products do not dispel or deter the much needed research
investment in the area.

Certain specific themes have been selected for consideration by the
conference—soil structure, waste recycling, residue management, the
rhizosphere, nitrogen fixation, mycorrhizae and biological control. It should
be stated that the keynote speakers were not chosen for their commitment to
any particular form of agriculture. Some of them have worked in and have
considerable experience of sustainable systems; others have not. It is important

to stress that the conference was not, and was never conceived as, a dogmatic idealogical platform for biological farming. Though our speakers may not have worked in or even considered this alternative before, their knowledge and experience of their respective subjects is of immense value to those of us—scientists and practitioners alike—who wish to promote more active research and development for alternative farming systems.

Initially we shall be considering the fate of organic residues and their management. What should be done with them and how best to recycle them within the sustainable agricultural system? This, and many other areas I shall mention, were highlighted in a summary of a conference in the USA which examined the approach and problems of the organic farming system (ASA, 1984). There is a great ignorance of the relationships that exist between soil and plant nutrients when residue management systems are altered. Dr. Parr considers the value of composting for the recycling of such waste materials. Composting and the use of compost is often seen to be at the theoretical heart of the biological agricultural approach with its capacity to improve soil physical properties, increase soil organic matter levels and provide a reservoir of slow release nutrients for plant growth. How much compost, however, should a biological farmer apply? On present experience if he resides in Europe perhaps 10–30 tonnes per hectare, if in China possibly up to 100 tonnes—such is the range quoted in the general literature.

Anyone who has made even the simplest perusal of the organic/biological literature will have noticed the emphasis placed on traditional crop rotations, principally involving legumes, and the inevitable reference to mycorrhizae. Both symbiotic topics are examined in detail in the papers by Drs. Sprent, Patriquin and Mosse. The study of nitrogen fixation, symbiotic and non-symbiotic, is an example of one area where considerable world research investment has taken place, although it is a biological system that is suppressed in the presence of synthetic nitrogen fertilizer additions. Much has been achieved in terms of improved inoculant strains of *Rhizobium* and a general understanding of the mechanism of biological nitrogen fixation. We now have a greater awareness also of the associative symbioses such as *Azospirillum*—perhaps others remain to be discovered. In addition to the unequivocal role of prokaryotes in nitrogen fixation is the important role of the soil microflora in the subsequent cycling of nitrogen through the ecosystem. We are indeed fortunate that we have an account of Dr. Patriquin's research into the cycling of nitrogen within a biologically manipulated farming system. Problems that arise here for the biological farmer are the fate and rate of release of nitrogen fixed in the legume residues. Can the nitrogen be released at the right time to synchronise with the requirements of the following crop? Such evidence that does exist suggests that the soils for the different farming systems may develop different biological properties and crop production capabilities (ASA, 1984).

The mycorrhizae have long held a respected position in the organic literature particularly as emphasised and propounded in the writings of Sir Albert Howard (Howard, 1941) resulting from his experiences during the first four decades of this century. The mycorrhizae were then envisaged as the link between the plant and the humus. Our knowledge and understanding of the biology of the mycorrhizae, though severely hampered by our inability to culture them axenically in the laboratory, has increased greatly in the last two decades, not least through the major contributions made by one of our keynote speakers, Dr. Mosse. It is my experience that in the biological agricultural fraternity the role of mycorrhizal fungi and their benefits to the plant is poorly understood and often overstated, perhaps as a legacy of the earlier writers. That mycorrhizal fungi assist in the uptake of phosphate (Tinker, 1980) and trace metals (Tinker, 1978; Timmer & Leyden, 1980), and possibly influence water and nutrient status via hormonal influences (Allen *et al.*, 1982), is not in great dispute. They appear to have a major manipulative role to play in the infertile soils commonly found in tropical regions and have been proposed as a promising technology for subsistence farmers in such regimes (Sanchez & Salinas, 1981). Are inoculants, however, the only route to enhance their activity or can the biological farmer so manage the ecosystem as to exploit their full potential? That phosphate efficiency can be improved in such systems has received some initial support from research at Pullman, Washington (cited in Elliott *et al.*, 1984) which indicated that a soil farmed organically since 1910 had significantly higher available P levels than an adjoining conventional farm. The organic farm had received minimal inputs in the form of manure and nutrients from outside. Does this reflect differences in actual biological activity or is it a reflection of the composite sustainable approach with its fundamental rotations and improved erosion control.

Improvements in soil management techniques and manipulations to increase soil fertility by maintaining soil organic matter levels are basic to the biological approach. Within this philosophy is the often-stated maxim that a healthy soil is full of life—both of a macro- and a microbiological nature. Microbial activity is particularly intense around the root regions of plants and, although the rhizosphere is a concept that undergoes periodic shifts of definition and delineation, its actual role and effects on the host plant are still obscure. It can be portrayed as being both beneficial and detrimental (Brown, 1975), given the range and diversity of microorganisms that are able to colonise or utilise the substrates emanating from the root region. One potential benefit of the mere presence of such a microflora is its possible protective effect on the host roots from pathogenic microorganisms providing an intensely active and therefore competitive environment to overcome. This links the rhizosphere directly with biological control, another theme of the conference. A recent addition in this area has been the concept of the 'quasipathogen'—microorganisms closely associated with the root regions that are not overtly pathogenic but may cause unseen, limited damage that cumulatively leads to considerable reduction in growth due to the divergence of plant energy to replace damaged roots or root tips. Both Drs. Lynch and Cooke examine the rhizosphere and biological control and place them in the context of sustainable approaches. In passing it is interesting to note that the medical world was not slow to recognise the importance, and exploit the value, of Dr. Waksman's soil-inhabiting actinomycetes. Perhaps one day these organisms will be manipulated, exploited and targeted by the plant protection community at the enemies at home—the plant pathogens in the soil.

An important feature of all the themes that I have mentioned is the interactions that must occur between them. The conference was deliberately designed to have only one session running so that all—rhizobiologists, soil scientists, plant pathologists, nitrogen fixologists, etc.—could appreciate how these interactions are important to an understanding and development of a sustainable farming ecosystem. Indeed, it was most encouraging that so many people travelled from far and wide to subject themselves to possibly unfamiliar if not outrageously new topics in an age of ever increasing reductionism.

I hope that this brief introduction to the conference has provided an insight, however meagre, into the sustainable approach—subsequent papers, I am sure, will reinforce, extend and correct what I have attempted to transmit. In order to explain my own enthusiasm for the subject, I should like to refer participants to the only major economic and field study to compare actual organic farms with matched conventional farms (Lockeretz *et al.*, 1984). Over a five year study period, maize (the major crop on both farm types) yields were 8% lower on organic farms. For soyabeans the difference was 5%. The greatest difference (43%) was found with winter wheat, a minor crop on both kinds of farms. If we consider at the same time the amount of private and public monies that have been poured worldwide, since the second world war, into what we now term "conventional agriculture" then these limited differences are astonishing. What might we achieve within biological sustainable approaches with similar massive financial resources?

Biological agriculture is a complicated and complex ecosystem to study. It is not a primitive and backward system of agriculture. It requires, even demands, a much greater understanding of basic soil microbial processes and plant/microbial interactions. The development of such farming systems will require long-term interdisciplinary research and the microbiologist will be at the centre of the wheel—this is the fundamental and relevant challenge that faces agricultural microbiologists in the future.

References

A.S.A. (1984). *Organic Farming: Current Technology and its Role in a Sustainable Agriculture.* American Society of Agronomy Special Publication No. 46 (D.F. Bezdicek & J.F. Power, eds.), Madison; Wisconsin.

Allen, M.F., Moore, T.S. & Chistensen, M. (1982). Phytohormonal changes in *Boutelova gracilis* infected by vesicular arbuscular mycorrhiza. II. Altered levels of gibberelin like substances and abscisic acid in the host plant. *Canadian Journal of Biology*, **60**, 468–477.

Boeringa, R. ed. (1980). Alternative methods of agriculture. Description, evaluation and recommendations for research. *Agriculture and Environment*, **5**, 1–199.

Besson, J-M. & Vogtmann, H. eds. (1978). *Towards a Sustainable Agriculture.* 1st. IFOAM International Conference, Sissach Switzerland, 1977. Verlag Wirz; Aarau.

Brown, M.E. (1975). Rhizosphere microorganisms—opportunists, bandits or benefactors. In *Soil Microbiology* (N. Walker, ed.), pp. 21–38. Butterworths; London.

Edens, T.C., Frigden, Cynthia & Battenfield, Susan L. eds. (1985). *Sustainable Agriculture & Integrated Farming System.* Michigan State University Press; East Lansing.

Elliott, L.F., Papendick, R.I. & Parr, J.F. (1984). Summary of the organic farming symposium. In *Organic Farming: Current Technology and its Role in a Sustainable Agriculture* (D.F. Bezdicek & J.F. Power, eds), pp. 187–192. American Society of Agronomy; Madison, Wisconsin.

Friend, G. (1983). The potential of an alternative agriculture. In *Sustainable Food Systems* (D. Knorr, ed.), pp. 28–47. Ellis Horwood; Chichester.

Hill, S. & Ott, P. eds. (1982). *Basic Technics in Ecological Farming* and *The Maintenance of Soil Fertility.* Proceedings of the 2nd and 3rd IFOAM International Conferences, Montreal and Bruxelles. Birkhauser Verlag; Basel.

Howard, A. (1941). *An Agricultural Testament.* Oxford University Press; London.

Lockeretz, W. ed. (1983). *Environmentally Sound Agriculture.* Praeger Publishers; New York.

Lockeretz, W., Shearer, L., Kohl, D. & Klepper, R.W. (1984). Comparison of organic and conventional farming in the corn belt. In *Organic Farming: Current Technology and its Role in a Sustainable Agriculture* (D.F. Bezdicek & J.F. Power, eds.), pp. 37–48. American Society of Agronomy; Madison, Wisconsin.

Merrill, Margaret C. (1983). Eco-agriculture: a review of its history and philosophy. *Biological Agriculture & Horticulture*, **1**, 181–210.

Sanchez, J.A. & Salinas, J.G. (1981). Low-input technology for managing oxisols and ultisols in tropical America. *Advances in Agronomy*, **34**, 279–406.

Stonehouse, B. ed. (1981). *Biological Husbandry.* Butterworths; London.

Timmer, L.W. & Leyden, R.F. (1980). The relationship of mycorrhizal infection to phosphorus induced Cu deficiency in sour orange seedlings. *New Phytologist*, **85**, 15–23.

Tinker, P.B. (1978). Effect of vesicular arbuscular mycorrhiza on plant nutrition and growth. *Physiologie Végétale*, **16**, 743–751.

Tinker, P.B. (1980). Root soil interactions in crop plants. In *Soil and Agriculture* (P.B. Tinker, ed.), pp. 1–34. Critical Reports on Applied Chemistry No. 2. Blackwell; Oxford.

U.S.D.A. (1980). *Report and Recommendations on Organic Farming.* U.S. Government Printing Office; Washington, D.C.

The Microbiology of Soil Structure

Richard G. Burns and Julie A. Davies

Biological Laboratory, University of Kent, Canterbury, Kent, CT2 7NJ, U.K.

INTRODUCTION

Soil aggregates are agglomerations of sand, silt, clay and organic matter arbitrarily described as having a diameter of between 0.25 and 10 mm. Aggregates are themselves composed of clusters of clay crystals $<50\mu$m called domains, and of silt-size fractions bonded by organic materials called microaggregates (see Edwards & Bremner, 1967; Emerson, 1977; and Oades, 1984 for definitions). Domains, microaggregates and aggregates represent stages in the genesis of a soil's structure and have rather different physical, chemical and biological mechanisms involved in their formation.

Fertile soils have a high proportion of their components associated in stable aggregates. This stability is principally to incoherence caused by rainwater and irrigation (aggregates are said to be water stable), but also to wind erosion and compaction by agricultural machinery. When unstable aggregates are wetted they disintegrate because they are not strong enough to withstand the pressures of entrapped air and unequal swelling. This process is called slaking (Emerson, 1967). The liberated microaggregates may disintegrate further as clay particles are detached. Not surprisingly slaking and clay dispersal have a detrimental effect on soil fertility (Shanmuganathan & Oades, 1982).

Aggregates may be measured for their stability to water by wet-sieving, compaction or dispersion. Resistance to dry-sieving and ultrasound may also be a useful guide to soil structure. In addition, the strength and nature of the mechanisms that bind soil particles together can be investigated using organic solvents (e.g. ethanol), mineral acids (e.g. HCl) and solutions of cations (e.g. Na^{2+}). The degree of aggregate disruption after a given time of exposure can be accurately quantified turbidimetrically or by weighing the dispersed components. A further method of assessing aggregation arises from the observation that stable aggregates have a stable and extensive pore structure. The volume of pores can be measured using gas diffusion rates

(Currie, 1983) and it is quite obvious that the dimensions and continuity of aggregate pores have a major effect on soil structure.

A degree of caution is needed when interpreting soil stability data, firstly because the different ways of measuring stability do not relate well to each other, and secondly because stability in the laboratory is not necessarily synonymous with stability in the field.

Well-aggregated soils, in contrast with unstable soils, show increased plant root penetration and anchorage, improved aeration and drainage, more rapid water and solute diffusion, and higher ion exchange capacities and humic contents. Furthermore, stable soils are less likely to compact, crust, crack and erode than are unstable soils. In general, stable soils also support a large and varied microbial biomass.

Because of the well-established link between soil structure and fertility (Baver, 1968; Lynch, 1984), the many microbial factors which influence soil aggregate formation and stabilization have been a major research interest for almost fifty years. Some of these microbiological factors are discussed in this chapter.

MICROORGANISMS AND AGGREGATE FORMATION

There is a distinct relationship between the organic matter content of a soil and its degree of aggregate stability (Baver, 1956; Low, 1962; Aringhieri & Sequi, 1978; Chaney & Swift, 1984). However, it has long been recognised that the mere addition of organic matter to soil is of little immediate benefit to soil structure. Moreover, even the prolonged incubation of a sterile soil to which organic matter has been added produces little in the way of increased aggregation (Martin & Waksman, 1940; Peele, 1940). In fact, it is the subsequent decomposition of the organic amendment that leads to structural improvements (Browning & Milan, 1941; Miller & Kemper, 1962; Tisdall *et al.*, 1978). Tisdall *et al.* (1978) used four organic amendments (oaten hay, sheep faeces, cellulose, cellulose+N) at 2% and 4% w/w. All increased aggregate stability within 1–4 weeks and the stability was retained for up to 32 weeks. Addition of easily degraded carbon sources, e.g. glucose, results in shorter lag phase and a more rapid (1–4 days) increase in stability followed by a more sudden decline (Harris *et al.*, 1966a). It is obvious from these common observations that the improvement in soil structure is due to the flush of microbial biomass, an increase in microbial metabolites, or some physico-chemical change brought about by microbial activity.

Filamentous Structures

Intuitively it would seem that filamentous fungi and actinomycetes must play an important part in enmeshing and entrapping soil particles to form aggregates. Indeed, early studies tended to endorse the aggregating properties of fungi although the hyphae were quite rapidly destroyed by colonizing bacteria with a concomitant decline in stability (McCalla, 1943; 1945; 1947; 1950; Martin & Anderson, 1943; Swaby, 1949). Other studies (e.g. McGill *et al.*, 1973; Low & Stuart, 1974) have tended to confirm the view that, although periodic organic additions may stimulate the growth of many filamentous organisms, these are generally ephemeral structures. As a consequence aggregates formed by the activities of filamentous fungi must rely on other mechanisms if they are to retain their improved structure. Notwithstanding, some hyphae are more persistent than others—possibly because they are in inaccessible microsites or because they have melanic cell walls with associated bacteriostatic properties. Bond (1960) and Bond & Harris (1964) reported instances of soils which appeared to have a prolonged dependence on fungal hyphae for stability. In contrast, shorter and narrower projections such as the pili and flagella of bacteria are unlikely to have a significant direct effect on aggregation. However, inasmuch as these structures may help bacteria traverse a zone of repulsion and establish themselves at a soil surface, motile bacteria may be important in polysaccharide production once clay- or silt-microbe contact has been achieved (Marshall, 1976; Burns, 1979, 1980). Thus there may be a correlation between appendaged bacteria and aggregation but it is not due to the physical entrapment of soil particles.

Most studies of filamentous entrapment concern free-living fungi; fungi in symbiotic mycorrhizal associations or merely in close proximity to a plant root present a rather more confused picture. Indeed, it is possible to see a major evolutionary thrust in land plants as favouring those species which stimulate the development of a rhizosphere environment which, in turn, brings about a localised improvement in soil structure. Thus those plants that were capable of initiating this change (partly through inducing the growth of filamentous and polysaccharidic microorganisms) were at a competitive advantage compared with those that could not. Given the nutrient insufficiency of non-rhizosphere soil, the rhizosphere is almost certainly the principal location of microbe-induced aggregation.

Stabilization of aggregates in the rhizosphere has been reported frequently (Sutton & Sheppard, 1976; Tisdall & Oades, 1979) and is often due to vesicular arbuscular mycorrhizae (VAM). The mycorrhizal species *Glomus* may play an important role in stabilizing sand dunes (Koske *et al.*, 1975). Foster (1981) examined sand dune aggregation using transmission electron microscopy and observed a change in emphasis from bacterial- to fungal-induced aggregation following the growth of the sand grass *Agropyron junciforme*. The mycorrhizal hyphae of *Glomus fasciculatus* and a *Penicillium* species were seen to be entangling the sand grains. In contrast, Reid & Goss (1981) did not find such a correlation between VAM and aggregate stability. Notwithstanding, the aggregating qualities of filamentous rhizosphere microorganisms, whether they are mycorrhizal or free-living, are broadly accepted even though it is not always obvious where the direct effect of root hairs ends and microbial entanglement begins. A further complication is that plant root polysaccharides have aggregating properties of their own and indeed mucigel at the root surface (Oades, 1978) is frequently seen embedded with clay particles (Foster & Rovira, 1976; Guckhert *et al.*, 1975).

The overlapping and difficult-to-resolve contributions of filamentous structures and plant and microbial gums was illustrated by Clough & Sutton (1978) who studied the aggregating properties of *Glomus* in association with bean (*Phaseolus vulgaris*) and grass (*Calamovilfa longifolia, Andropogon* sp.) plants. Light and scanning electron microscopy certainly revealed that fungal hyphae were associated with sand grains. In addition, however, an amorphous polysaccharide material, apparently acting as an adhesive, was seen between the sand grains and the hyphae. Earlier research with a free-living fungus *Cladosporium* (Martin, 1945) had suggested that the contributions of mechanical and adhesive mechanisms were approximately equal although extracellular adhesins dominated bacterial binding. Aspiras *et al.* (1971a) chose a variety of fungi and actinomycetes to investigate the relative importance of physical entrapment and adhesive binding. Using sonication to disrupt the filamentous structures they concluded that the mucilage on the hyphae was of paramount importance in stabilizing aggregates. However, recent work by Carter & Oades (cited by Oades, 1985) suggests that the binding action of fungal hyphae may have been underestimated. These researchers used periodate treatment to destroy polysaccharides in aggregates and found no decline in aggregate stability. This underlines an earlier observation that the stability of aggregates was directly related to the extent of hyphal development within those aggregates (Tisdall & Oades, 1980a).

These apparently contradictory reports may be rationalised if mycelial structures and mucilages are involved in different stages in aggregate formation and stabilization. Indeed, filamentous organisms have been implicated by some primarily in the formation of aggregates (Hubbell & Chapman, 1946; Harris *et al.*, 1966b; Tisdall & Oades, 1980b; 1982) and thus may be concentrated in the larger pores. Polysaccharides may be more important in the process of cementing together clay particles and domains to form microaggregates. However, it is almost certainly unwise to search for a unified concept of soil aggregation: no doubt entrapment and adhesion are both important but their relative contributions will vary from soil to soil, with time and climate, depending on plant cover and agricultural treatment, but above all, according to which microbial species are involved.

Microbial Polymers

A number of microbial metabolites have been implicated in soil stabilization, for example lipids, proteins and phenolics, but by far the greatest emphasis has been placed on extracellular polysaccharides. There is much evidence to support the belief that polysaccharides play a major role in soil aggregation not least because of the large number of soil microorganisms capable of synthesising exopolysaccharides, at least *in vitro* (Hepper, 1975). What has come to be known as the 'polysaccharide theory' of soil aggregation was for a long while a subject of controversy and debate. In retrospect, it seems strange that there should have been a controversy at all but it probably arose from the dogmatic and contrasting viewpoints that either polysaccharides were always of major importance or that polysaccharides were of limited and only occasional importance. Of course, the truth is somewhere between the two, namely microbial extracellular polysaccharides have an important influence on aggregate stabilization in some soils under certain conditions. Nevertheless, it is worth summarising the arguments used by the supporters and opponents of 'polysaccharide theory'.

Aggregate stability is positively correlated with soil polysaccharide content (Rennie *et al.*, 1954; Santanatoglia & Fernandez, 1983) yet Webber (1965) monitored changes in the polysaccharide content of a clay soil over a four month growing season and found no relationship between degree of aggregation and polysaccharide content. Interestingly, Mehta *et al.* (1960) had suggested that well-aggregated soils encourage the growth of polysaccharide producing microorganisms thus rendering redundant questions about the involvement of polysaccharides in aggregate initiation.

Destruction of soil polysaccharide by an oxidant such as sodium periodate will often cause a decline in aggregate stability (Greenland *et al.*, 1962; Clapp & Emerson, 1965). However, Mehta *et al.* (1960) showed that natural aggregates from a brown earth forest soil were unaffected by periodate treatment although synthetic polysaccharide-containing aggregates were destroyed. The work of Greenland *et al.*, (1962) seemed to suggest that crop sequences had an important bearing on the contribution of polysaccharides: a rotation with only 2 years of pasture contained aggregates that were very sensitive to periodate treatment whereas aggregates under permanent pasture were highly resistant. Tisdall & Oades (1980) again differentiated between microaggregates and aggregates and reported that aggregates >50 μm diam. were unaffected by periodate treatment. As mentioned earlier, they concluded that polysaccharides are most important in stabilizing microaggregates and domains.

Some of the work using periodate to destroy cementing polysaccharides (e.g. Stefanson, 1971; Tisdall & Oades, 1979; Hamblin, 1980) is of rather questionable value given the observations of Cheshire (1974 & 1983). In fact it may be necessary to subject microaggregates to prolonged exposure (>150h) to excess sodium periodate and tetraborate in order to destroy most of the polysaccharide and bring about dispersal. Cheshire (1983) reported a linear relationship between residual carbohydrate (after various periods of oxidation) and aggregate stability. A subsequent study (Cheshire, 1984) using a number of pasture and arable soils under different types of cultivation confirmed the relationship between residual microbial carbohydrate and aggregate stability in 13 of 15 soils examined. It is possible that there are at least two types or associations of polysaccharides involved in aggregation; one type that is easily destroyed by oxidants and another that is highly resistant. Even though Oades (1984) has argued that extreme periodate treatment is rather non-selective and destroys not only the binding polysaccharides but also other agents of aggregation, it is possible that rather more severe oxidation is needed to disrupt clay-polysaccharide associations than had been thought previously. Certainly polysaccharides attached to smectite clays such as montmorillonite are difficult to destroy with periodate (Olness & Clapp, 1975).

Microbial polysaccharide isolated from cultures of soil microorganisms and added to soil or incorporated into synthetic soil aggregates improves stabilty (Rennie *et al.*, 1954; Martin & Richards, 1963; Moavad *et al.*, 1976; Martin, 1971). Unfortunately any improvement in aggregation is usually short-lived although this has not stopped investigations into the use of microbial polysaccharides as soil conditioners (see page 104).

Acceptance of a crucial role for polysaccharides in soil aggregation depends not only on the experimental data but also on the resolution of two major conceptual problems. Firstly, in carbon-limited environments such as soil, extracellular polysaccharides should be biodegraded rapidly; and secondly exopolysaccharide production is generally a consequence of overflow metabolism: in other words a response to nutrient excess and high aeration conditions uncommon in soil.

A number of explanations for polysaccharide persistence in soil have been suggested (Martin, 1971; Burns, 1982). Some microbial polysaccharides are intrinsically difficult to degrade because they are heteropolymers (i.e. contain two or more different sugar residues), are branched rather than linear, or have complex fibrillar structures. As a result a suite of enzymes acting in concert and probably arising from different microbial species may be necessary for the destruction of adhesive polysaccharides. Cheshire *et al.* (1974) have cast doubt on the notion that soil binding agents are inherently resistant to biodegradation by extracting polysaccharides from soil, adding them back to soil and recording their rapid disappearance. In a comparable period little of the polysaccharide in undisturbed aggregates was decomposed. This work suggested that any persistence of polysaccharides in soil must be due to their inaccessibility or an association with inorganic or other organic moieties which is broken during extraction. It is equally possible, however, that the prolonged extraction technique involving NaOH, HCl, centrifugation and filtration altered the structure and properties of the native polymer. Nevertheless, it is true that, for instance, 1–3 linkages are more recalcitrant than 1–4 linkages and fungal polysaccharides are, in general, more stable than those of bacterial origin. A possibly relevant observation also made by Cheshire (1984) was that polysaccharides containing glucose, arabinose and xylose were more resistant to periodate oxidation than those composed of galactose, mannose, fucose and rhamnose. It is not known whether this distinction applies to the biodegradation of these polysaccharides.

A second possible explanation for polysaccharide stability is that, because they are primarily concerned with binding microaggregates, they are located in microsites which are inaccessible to many microorganisms (Cheshire, 1974). In fact it has been estimated from considerations of pore volumes and microbial numbers that $>90\%$ of the surfaces in soil are protected from microbes and their enzymes (Adu & Oades, 1978b). Furthermore, less than 30% of the bacterial population produce polysaccharases (Cheshire *et al.*, 1974). Polysaccharides may also be protected from attack by intercalation within clay lattices or because the bonds which are most susceptible to cleavage are masked by surface interactions. Another factor is that charged clay particles have a high affinity for enzyme protein and adsorption usually results in irreversable inactivation of any scavenging polysaccharases.

Adu & Oades (1978a) measured CO_2 release from ^{14}C-labelled glucose or starch added to synthetic aggregates (either before or after moulding) and natural aggregates. The distribution of substrate in synthetic aggregates was considered to be in both small and large pores whilst that in natural aggregates only penetrated the larger pores. They convincingly demonstrated that polymeric starch but not monomeric glucose was protected from biodegradation due to its inaccessibility in micropores. Surprisingly, perhaps, this protection was more pronounced in sandy loam than in clay aggregates.

The use of electron microscopy to study soil microenvironments (Faull & Campbell, 1979; Kilbertus, 1980; Foster, 1981; Foster & Martin, 1981) has shown that mucilages, either attached to microorganisms or remote from them, become coated by clay platelets effectively restricting the access of degradative microbes and enzymes. It is also apparent (Oades, 1984) that polysaccharides are not highly mobile in soil, and domains and microaggregates are formed by deposition of clay and silt from the soil solution rather than by the diffusion of exposed polymers.

Soil polysaccharides are known to form complexes with metal ions and polyphenolic humic materials. Both these interactions will result in an increase in persistence for the polysaccharide. In general, the more aromatic and less soluble humic acids are better stabilizing agents than fulvic acids although there are some exceptions (Martin *et al.*, 1978). Cu, Fe, Zn and Ca will all reduce the decomposition rates of polysaccharides (Martin *et al.*, 1966). The separate influence of cations and anions on clay flocculation (and thus the initial stages of aggregate formation) is outside the purview of this chapter.

The possibility exists that, at least in rhizosphere soils, aggregate formation and degeneration is in a steady state and that soils gain cementing polysaccharides at the same rate as they lose them. The notion of a balance between microbial synthesis and degradation of adhesive agents dependent upon carbon and energy sources is implicit in the work of Stallings (1953). Certainly aggregate stability in bare fallow soils, with little organic input, declines rapidly although this may be due as much to hyphal disintegration as to polysaccharide loss (Tisdall & Oades, 1982).

Finally, restrictions in oxygen diffusion (Tiedje *et al.*, 1984) and unsuitable C:N ratios within aggregate microenvironments will undoubtedly retard carbohydrate mineralization.

The polysaccharides involved in soil aggregation are predominantly microbial in origin. This is known because microbial biopolymers contain galactose, glucose and mannose and much smaller quantities of the plant saccharides arabinose and xylose (Cheshire, 1977, 1979). Thus the ratio of galactose and mannose to arabinose and xylose (G+M:A+X) is high ($>$2.0) for microbial polysaccharides and low ($<$0.5) for plant polysaccharides. The microbial mucilages will also contain polyuronic acids and a host of amino compounds (Swincer *et al.*, 1968).

Particle size analysis (Turchenek & Oades, 1978, 1979; Tiessen & Stewart, 1983) indicates that recalcitrant microbial polysaccharides (and organic C in general) are ultimately concentrated in the stable silt-size fractions which they have helped to form. Larger particles have lower G+M:A+X ratios and may thus contain a higher proportion of plant polysaccharides. Indeed, from this and the work of Anderson *et al.* (1981), it is apparent that the various size fractions of aggregates have very different organic carbon (and nitrogen and sulphur) compositions and that different organics may consistently be involved in the adhesion of fine and coarse clays compared with those in fine and coarse silts. As a consequence, the construction and composition of even the microaggregates may be far more complicated than is realised.

The adsorption of polysaccharides and other organics to clays, which involves a multitude of physical and chemical mechanisms including van der Waals forces, hydrogen bonding, polyvalent cation bridging (especially Al, Fe, Ca and Mg) and ion exchange, has been discussed by Martin (1971), Hayes & Swift (1978), Theng (1983) and Oades (1984). Murray & Quirk (1985) have reviewed in detail the interparticle forces involved in aggregate stabilization.

IMPROVEMENT OF SOIL STABILITY

Modern intensive agriculture has in many instances, led to deterioration of soil structure with consequential declines in crop yield. In addition many marginal soils with inherently poor aggregate stability must now be considered for crop production. As a result there has been much research aimed at identifying the changes that occur during disaggregation and at recovering or improving soils for agricultural use.

Three principal ways of using microorganisms to improve aggregation have been investigated: addition of organic matter to stimulate aggregate-forming microorganisms; addition of microbial gums; and the use of microbial inocula.

Organic matter amendment

One method of improving soil structure is to stimulate the indigenous microflora in the hope that metabolites capable of initiating and stabilizing aggregates will also be produced. Indeed, the age-old agricultural practice of incorporating organic matter into soil not only returns plant nutrients but also improves 'tilth'—a direct consequence of aggregate formation. Thus the addition of organic matter to soil is a comparatively easy and non-controversial (unless sewage sludge containing heavy metals is involved) treatment. Some of the early studies (e.g. Martin, 1942, 1945; Low, 1954; Rennie *et al.*, 1954; Griffiths & Jones, 1965) demonstrated the value of adding easily-degraded carbon sources to soils or to artificial aggregates. After a brief lag phase the flush of microbial growth and production of polysaccharide metabolites gave rise to an increase in aggregate stability. Usually, however, the improved stabilization is ephemeral and destabilization takes place as the amended substrate is exhausted and the microbial population returns to its original level. A few field experiments implicate humic materials in the long term stabilization of a carbon/microbial flush/polysaccharide-induced improvement in stability. For example Griffiths & Burns (1972) showed that glucose added to field plots produced the predicted improvement in aggregation but the subsequent rapid decline in stability could only be arrested by application of a phenolic compound such as tannic acid. These authors suggested that tannic acid 'protected' the new polysaccharide in aggregates from rapid biodegradation. These data were consistent with those arising from *in vitro* experiments in which polysaccharide-amended artificial aggregates were given long term stability by tannic acid or the phenolic constituents of decomposing herbage.

In recent years the controlled addition of plant and animal residues to soil has been discussed (Eiland, 1981; Lynch, 1984) not only in the context of improving nutrient levels and soil structures but also as a solution to environmental problems arising from organic waste disposal. However, the economic and biological details of such additions are still largely a mystery and need further evaluation. That there are actual disadvantages to plant growth of organic matter additions is evident from the work of Lynch and colleagues (Lynch, 1977; Lynch & Elliott, 1983; Lynch *et al.*, 1980).

Addition of microbial polysaccharides and other metabolites

Many laboratory studies confirm the important role of polysaccharides in aggregate formation and experiments in which microbial polysaccarides have been added to soil are numerous (see, for example, Jones & Griffiths, 1967; Griffiths & Burns, 1972; Lynch, 1981 and p. 100). However, the extensive application of microbial polymers to field plots has not been attempted because of the hitherto prohibitive expense of these compounds. In recent years, the large scale production of biopolymers for use in the petrochemical industry has led to cheaper polysaccharides and more efficient microbial producers. Notwithstanding, most of the studies of soil conditioners has involved synethic polymers such as polyvinyl acetate, polyacrylonitrile and polyethylene glycol (for reviews of the work see Hayes, 1980; De Boodt, 1985). Cellulose xanthate has attracted interest in this context (Menefee & Hautala, 1978). Recently we have begun to screen a range of microbial polysaccharides as possible soil conditioners (Davies & Burns, unpublished observations). We have found that a number of easily- and cheaply-produced polysaccharides (e.g. xanthan, dextran, succinoglycan, pullulan, curdlan) are effective soil stabilizers in the short term. Prolonged incubation of amended soils leads to a rapid decline in aggregate integrity. We are currently investigating ways of complexing polysaccharides to enhance stability—in a manner analogous to enzyme-polyphenol stabilization (Sarkar & Burns, 1984), consistent with the observations of Benoit & Starkey (1968), and broadly mimicing a common if not unanimous view of the process of aggregate formation *in situ*.

Microbial inocula

The effect of additions of microorganisms to soil may be studied by either direct application of cultures to soil or by incorporation of inocula into synthetic aggregates. Obviously the former is the only feasible possibility for field application. Any subsequent changes in aggregate stability may be due to one or more of a number of factors. The microbial inoculum may establish itself, proliferate, and contribute directly to aggregation, or alternatively, serve as a substrate for the indigenous microflora which will in turn induce improvements in stability. A third possibility is for the inoculum to contain a high quantity of associated polysaccharides or other adhesins before addition and thus cell viability and colonization becomes irrelevant. In other words the additive is merely a crude preparation of cells, extracellular polymers and spent medium. For many reports of improved aggregation the second and third factors are the most feasible. The first is less likely as there are a number of barriers to be overcome before colonization and establishment can occur.

Soil is an environment in which substrate levels fluctuate wildly but are generally growth limiting. As a consequence competition for the limited resources is fierce and in a mature or climax soil population all the components of the microflora are well-suited to the environment. Any introduced microorganisms have little chance of long term success because it is likely that most, if not all, of the available habitats are occupied and the resources utilized. Thus, unless the introduced organism can occupy a vacant niche it is unlikely to survive for long enough to make a direct contribution to the overall metabolism of the soil. This is why claims for the efficacy of 'microbial fertilizers', 'microbial activators' and 'microbial conditioners' are often viewed with scepticism. There are ways of overcoming these barriers to colonization but they will involve the addition of new or specific substrates along with the inoculum. Other topical possibilities include the use of genetically-modified microorganisms. Inocula of this type may have for example, the capacity to effectively utilize unusual substrates already present in soils but not readily metabolized (e.g. humic acid components) or be aggressive competitors because either they possess transport mechanisms with very high affinities for low concentrations of substrates or because they produce bacteriostatic agents that inhibit the indigenous microflora. No doubt these engineered microorganisms could be copious polysaccharide producers or capable of extensive hyphal development. However, if we assume that the simultaneous addition of organic carbon and nitrogen is at present uneconomic, then, in the absence of these 'super-bugs', the use of heterotrophic inocula on a field scale is precluded. However, recent studies using photosynthetic microorganisms as soil additives are encouraging. Phototrophs will not, by definition, compete for soil carbon or even nitrogen. A few papers which specifically concentrate on the addition to or incorporation of named microbial species into natural or artificial aggregates are discussed here.

An exemplary study by Aspiras *et al.* (1971a) revealed that different microbial species in the presence of an additional carbon source (sucrose) produced different chemical and physical binding agents. Employing various extractants and oxidants (e.g. $NaIO_4$, $Na_4P_2O_7$, hexane, propanol, benzene) these workers showed that *Agrobacterium radiobacter, A. tumefaciens* and *Rhizobium trifolii*, added to previously sterilized silt loam and silty clay loam artifical aggregates, caused binding by polysaccharides. Stabilization by *Bacillus polymyxa* and a number of *Streptomyces* species was attributable to a combination of polysaccharides and humic substances. For many species of fungi (e.g. *Altenaria tenuis, Stachybotrys atra, Penicillium* spp. and *Aspergillus* spp.) humic and lignin-like materials were the dominant aggregating agents. Alone of those tested, *Mucor heimalis* produced waxes and fats which served as stabilizing compounds. Aspiras *et al.* (1971a) also measured the resistance of aggregates to physical disruption using ultrasound. This indicated the additional importance of filamentous structures produced by fungi and actinomycetes—although it may be the binding materials associated with the mycelium rather than the filaments themselves that are important. Especially effective in mechanical binding were *A. tenuis* and *S. atra* in clay soils. Unfortunately no report was made of changes in microbial biomass during the incubation period prior to measurement of stability nor whether the autoclaved aggregates remained axenic. A second publication (Aspiras *et al.*, 1971b) using the same microbial inocula endorsed the importance of the concomitant addition of substrate: inoculated aggregates plus sucrose or corn stover were stable for much longer than those without the organic amendment.

In an earlier piece of work Harris *et al.* (1966) isolated ten fungal species and a number of bacterial spore formers (probably *Bacillus* spp.) from soil and then investigated their effects on addition to sterilized aggregates of the same soil. The fungi differed in their ability to effect aggregation depending on the presence of sucrose (2% w/w). The stabilization of aggregates >2mm coincided with the appearance of macroscopic mycelia. Improvements in the stability of aggregates amended with bacteria was not related to microbial numbers again pointing to the function of adhesive metabolites especially in the production of <2mm aggregates.

In another laboratory experiment Lynch (1981) found that adding 7–day cultures of *Azotobacter chroococcum*, *Lipomyces starkeyi* and a *Pseudomonas* sp. to a silt loam soil induced an immediate improvement in stability that was proportional to the number of cells and biomass (0–10mg dry weight (gsoil)$^{-1}$) added. *Mucor hiemalis*, in contrast, actually inhibited aggregation although the same fungus promoted stabilization in a clay soil. This experiment purposely concentrated on the direct effect of cell additions and not any subsequent effect due to inoculum establishment and proliferation. Thus the importance of chemical bonding agents is emphasised in contrast to any physical entrapment or the effect of new metabolites produced by growing cells.

A detailed study of microbial inocula and soil stability has been described in a series of papers by Dabek-Szreniawska (1974, 1977a, 1977b). *Bacillus* sp., *Arthrobacter* sp. and *Cytophaga* sp. were added (4mg wet wt (gsoil)$^{-1}$) to sterilized and non-sterilized loamy sand soil with or without additional carbon (1.5% glucose or cellulose). All these species were highly viscous (i.e. were producing copious amounts of polysaccharide) at the time of addition to soil. In general, carbon amendments stimulated aggregation and although the inoculated bacteria failed to establish themselves in the aggregate their rate of decline was much slower than in the non-carbon-amended aggregates. *Cytophaga* plus cellulose added to non-sterile soil produced the most marked and prolonged improvement in stability suggesting to the author that an interaction of the indigenous and introduced species was important for effective stabilization. As reported by others, no changes in stability occurred in the absence of an additional carbon source.

Gasperi-Mago & Troeh (1979) tested the effect of microbial additions on the stability of soils subjected to simulated rainfall. Sterile and non-sterile soil <3mm was sprayed with nutrient broth containing *Penicillium* sp., *Pseudomonas aeruginosa* and a *Streptomyces* sp. All microbial treatments improved stability after 15 days with the *Penicillium* sp. being the most effective.

Studies of photosynthetic soil additives are few in number even though they represent a promising avenue of research (see p. 105). The influence of microalgae on soil structure is described in a number of publications (Booth, 1941; Bond & Harris, 1964; Bailey *et al.*, 1973; Roychoudhury *et al.*, 1980) and it is well-known that some algal species produce copious amounts of extracellular polysaccharide *in vitro* (Lewin, 1956; Metting, 1981) and that in soil these polymers may be quite resistant to degradation (Verma & Martin, 1976). Lewin (1977) has described the aerial application of *Chlamydomonas mexicana* to field sites in S.W. USA at a rate of $1–5\,\mathrm{kg\,ha^{-1}}$. Depending on soil and climatic conditions the inoculum successfully colonized the soils resulting in biomass levels of $50–200\,\mathrm{kg\,ha^{-1}}$ after 3–4 weeks. In some instances signficant improvements in soil stability due to polysaccharide production occurred and, more importantly, increases in crop yield were recorded. Further studies of numerous algal species with the potential for improving soil stability are being carried out at the Scripps Institution of Oceanography in California (Lewin, 1983 and personal communication.)

Metting & Rayburn (1983) also used *Chlamydomonas mexicana* to treat sites in Washington State. Multiple applications over a number of years at rates of between 0.5×10^{10} and 3×10^{11} cells $\mathrm{ha^{-1}}$ (approximately equivalent to a surface application of $10^3\,\mathrm{g\,soil^{-1}}$) per application resulted in improved stability in all irrigated soils tested. Algae established themselves in the photosynthetic zone (2.6×10^8 cells g $\mathrm{soil^{-1}}$) and carbohydrate levels were increased—presumably due to the synthesis of binding polysaccharides and polyuronides. The possibility of using both carbon and nitrogen fixing phototrophs, such as some of the Cyanobacteria, has not yet been investigated in the context of soil aggregation.

CONCLUSIONS

Soil aggregates disintegrate under continuous cultivation for a host of reasons. For example, conventional tillage techniques improve aeration and stimulate the biodegradation of released binding materials; microbial filaments and root hairs are broken and stable soil pores are destroyed. In addition each year large volumes of well-aggregated soil are removed along with root crops, and the harvesting of plants inevitably reduces the amount of organic matter returned to the soil. Pesticides and fertilizer application, irrigation and burning of stubble may also have a detrimental effect on aggregation in some circumstances. As a consequence soil is constantly being eroded and unless this is arrested soil fertility declines (Biswas & Biswas, 1978).

Conventional good management of soil can counteract erosion and the value of rotations, grass leys, organic fertilizers and more recently minimum tillage, is recognised. Other less widely adopted methods, based on half a century of research into the complicated process of soil aggregation, involve the addition to soil of flocculating agents such as gypsum ($CaSO_4$), quicklime (CaO) and iron and aluminium salts. Since microorganisms and their products have a central role in aggregate stabilization it seems obvious that they could also be used to solve some of the problems. However, such is the complexity of their involvement that it has so far proved impossible to use microorganisms for the large scale improvement of soil in a way that would be both effective and economic. This review has outlined some of the avenues down which both fundamental and applied soil aggregation research is travelling. It is hoped that recent advances in biotechnology—especially with regard to the selection and genetic manipulation of microorganisms and their large scale growth in continuous culture—will supply strains which can be used in soil stabilization. The problem of soil erosion will not go away and must be solved before we lose more and more agricultural land to the desert.

References

Adu, J.K. & Oades, J.M. (1978a). Utilization of organic materials in soil aggregates by bacteria and fungi. *Soil Biology & Biochemistry,* **10**, 117–122.

Adu, J.K. & Oades, J.M. (1978b). Physical factors influencing decomposition of organic materials in soil aggregates. *Soil Biology & Biochemistry,* **10**, 109–115.

Anderson, D.W., Saggar, S., Bettany, J.E. & Stewart, J.W.B. (1981). Particle size fractions and their use in studies of soil organic matter. I. The nature and distribution forms of carbon, nitrogen and sulphur. *Soil Science Society of America Journal,* **45**, 767–772.

Aringheri, R. & Sequi, P. (1978). The arrangement of organic matter in a soil crumb. In *Modification of Soil Structure* (W.W. Emerson, R.D. Bond & A.L. Dexter, eds.), pp. 145–150. Wiley and Sons; Chichester.

Aspiras, R.B., Allen, O.N., Harris, R.F. & Chesters, G, (1971a). Aggregate stabilisation by filamentous microorganisms. *Soil Science,* **112**, 282–284.

Aspiras, R.B., Allen, O.N., Chesters, G. & Harris, R.F. (1971b). Chemical and physical stability of microbially stabilised aggregates. *Soil Science Society of America Proceedings,* **35**, 283–286.

Bailey, D., Mazurak, A.D. & Rosowski, J.R. (1973). Aggregation of soil particles by algae. *Journal of Phycology,* **9**, 99–101.

Baver, L.D. (1956). *Soil Physics,* 3rd. edn. Wiley; London.

Baver, L.D. (1968). The effect of organic matter on soil structure. *Pontificia Academia Scientiarum Scripta varia,* **32**, 383–413.

Benoit, R.E. & Starkey, R.L. (1968). Inhibition of decomposition of cellulose and some other carbohydrates by tannin. *Soil Science,* **105**, 291–296.

Biswas, M.R. & Biswas, A.K. (1978). Loss of productive soil. *International Journal of Environmental Studies,* **12**, 189–197.

Bond, R.D. (1960). The occurrence of microbial filaments and their effects on some soil properties. Divisional Report. C.S.I.R.O.; Adelaide.

Bond, R.D. & Harris, J.R. (1964). The influence of the microflora on physical properties of soils. I. Effects associated with filamentous algae and fungi. *Australian Journal of Soil Research,* **2**, 111–122.

Booth, W.E. (1941). Algae as pioneers in plant succession and their importance in erosion control. *Ecology,* **22**, 38–64.

Browning, G.M. & Milam, F.M. (1941). Rate of application of organic matter in relation to soil aggregation. *Soil Science Society of America Proceedings,* **6**, 96–97.

Burns, R.G. (1979). Interaction of microorganisms, their substrates and their products with soil surfaces. In *Adhesion of Microorganisms to Surfaces* (D.C. Ellwood, J. Melling & P.R. Rutter, eds.), pp. 109–138. Academic Press; London.

Burns, R.G. (1980). Microbial adhesion to soil surfaces: consequences for growth and enzyme activities. In *Microbial Adhesion to Surfaces* (R.C.W. Berkeley, J.M. Lynch, J. Melling, P.R. Rutter & B. Vincent, eds.), pp. 249–262. Ellis Horwood; Chichester.

Burns, R.G. (1982). Carbon mineralisation by mixed cultures. In *Microbial Interactions and Communities* (A.T. Bull & J.H. Slater, eds.), pp. 415–453. Academic Press; London.

Chaney, K. & Swift, R.S. (1984). The influence of organic matter on aggregate stability in some British soils. *Journal of Soil Science,* **35**, 223–230.

Cheshire, M.V. (1977). Origins and stability of soil polysaccharide. *Journal of Soil Science,* **28**, 1–10.

Cheshire, M.V. (1979). Nature and Origin of Carbohydrates in Soils. Academic Press, London.

Cheshire, M.V., Greaves, M.P. & Mundie, C.M. (1974). Decomposition of soil polysaccharide. *Journal of Soil Science,* **25**, 483–498.

Cheshire, M.V., Sparling, G.P. & Mundie, C.M. (1983). Effect of periodate treatment of soil carbohydrate constituents and soil aggregation. *Journal of Soil Science,* **34**, 105–112.

Cheshire, M.V., Sparling, G.P. & Mundie, C.M. (1984). Influence of soil type, crop and air drying on residual carbohydrate content and aggregate stability after treatment with periodate. *Plant & Soil,* **76**, 339–347.

Clapp, C.E. & Emerson, W.W. (1965). The effect of periodate oxidation on the strength of soil crumbs. *Soil Science Society of America Proceedings,* **29**, 127–130.

Clough, K.S. & Sutton, J.C. (1978). Direct observation of fungal aggregates in sand dune soil. *Canadian Journal of Microbiology,* **24**, 333–335.

Currie, J.A. (1983). Gas diffusion through soil crumbs: the effects of wetting and swelling. *Journal of Soil Science,* **34**, 217–232.

Dabek-Szreniawska, M. (1974). The influence of *Arthrobacter* sp. on the water stability of soil aggregates. *Polish Journal of Soil Science,* **7**, 169–179.

Dabek-Szreniawska, M. (1977a). The influence of *Arthrobacter* sp. on soil aggregation. *Zeszty problemowe Postepow Nauk rolniczich,* **197**, 319–328.

Dabek-Szreniawska, M. (1977b). the role of selected bacteria in the formation of water-stable aggregates independently of other microorganisms. *Zeszty problemowe Postepow Nauk rolniczich,* **197**, 339–354.

De Boodt, M. (1985). Application of polymers for stabilization of aggregates in agricultural soils. In: *Soil Colloids and Their Association in Soil Aggregates* (M.H.B. Haynes & M.F.L. De Boodt, eds.). NATO Publications, Springer Verlag (in press).

Edwards, A.P. & Bremner, J.M. (1967). Microaggregates in soils. *Journal of Soil Science,* **18**, 64–73.

Eiland, F. (1981). The effects of high doses of slurry and farmyard manure on microorganisms in soil. *Danish Journal of Plant and Soil Science,* **85**, 145–152.

Emerson, W.W. (1967). A classification of soil aggregates based on their coherence in water. *Australian Journal of Soil Research,* **5**, 47–57.

Emerson, W.W. (1977). Physical properties and structure. In *Soil Factors in Crop Production in Semi-Arid Environments* (J.S. Russell & E.L. Greacen, eds.), pp. 78–104. University of Queensland Press; Queensland.

Faull, J.L. & Campbell, R. (1979). Ultrastructure of the interaction between the take-all fungus and antagonistic bacteria. *Canadian Journal of Botany,* **57**, 1800–1808.

Forster, S.M. (1979). Microbial aggregation of sand in an embryo dune system. *Soil Biology & Biochemistry,* **11**, 537–543.

Foster, R.C. (1981). Polysaccharides in soil fabrics. *Science,* **214**, 665–667.

Foster, R.C. & Rovira, A.D. (1976). Ultrastructure of wheat rhizosphere. *New Phytologist,* **76**, 343–352.

Foster, R.C. & Martin, J.K. (1981). *In situ* analysis of soil components of biological origin. In *Soil Biochemistry* (E.A. Paul & J.N. Ladd, eds.), Vol. 5, pp. 75–111. Marcel Dekker; New York.

Gasperi-Mago, R.R. & Troeh, F.R. (1979). Microbial effects on soil erodibility. *Soil Science Society of America Journal,* **43**, 765–768.

Greenland, D.J., Lindstrom, G.R. & Quirk, J.P. (1962). Organic materials which stabilise natural soil aggregates. *Soil Science Society of America Proceedings,* **26**, 366–371.

Griffiths, E. & Jones, D. (1965). Microbiological aspects of soil structure. I. Relationships between organic amendments, microbial colonisation and changes in aggregate stability. *Plant & Soil,* **23**, 23–28.

Griffiths, E. & Burns, R.G. (1972). Interaction between phenolic substances and microbial polysaccharides in soil aggregation. *Plant & Soil,* **36**, 599–612.

Guckert, A., Chone, T. & Jacquin, F. (1975). Microplate et stabilité structurale des sols. *Revue d'Ecologie et de Biologie du Sol,* **12**, 211–223.

Hamblin, A.P. (1980). Changes in aggregate stability and associated organic matter properties after direct drilling and ploughing on some Australian soils. *Australian Journal of Soil Research,* **18**, 27–36.

Harris, R.F., Chester, G. & Allen, O.N. (1966a). Soil aggregate stabilisation by indigenous microflora as affected by temperature. *Soil Science Society of America Proceedings,* **30**, 205–210.

Harris, R.F., Chester, G. & Allen, O.N. (1966b). Dynamics of soil aggregation. *Advances in Agronomy,* **18**, 107–169.

Hayes, M.H.B. (1980). The role of natural and synthetic polymers in stabilising soil aggregates. In *Microbial Adhesion to Surfaces* (R.C.W. Berkeley, J.M. Lynch & J. Melling, eds.), pp. 263–296. Ellis Horwood; Chichester.

Hayes, M.H.B. & Swift, R.S. (1978). The chemistry of soil organic colloids. In *The Chemistry of Soil Constituents* (D.J. Greenland & M.H.B. Hayes, eds.), pp. 179–320. John Wiley; Chichester.

Hepper, C.M. (1975). Extracellular polysaccharides of soil bacteria. In *Soil Microbiology* (N. Walker, ed.), pp. 93–110. Butterworths; London.

Hubbell, D.S. & Chapman, J.E. (1946). The genesis of structure in 2 calcareous soils. *Soil Science,* **62**, 271–281.

Jones, D. & Griffiths, E. (1967). Microbial aspects of soil structure. II. Soil aggregation by the extracellular polysaccharide of *Lipomyces starkeyi. Plant & Soil,* **27**, 187–200.

Kilbertus, G. (1980). Etudes des microhabitats contenus dans les agrégats du sol. Leur relation avec la biomasse bactérienne et la taille des procaryotes présents. *Revue d'Ecologie et de Biologie du Sol,* **17**, 543–557.

Koske, R.E., Sutton, J.C. & Sheppard, B.R. (1975). Ecology of *Endogone* in Lake Huron and dunes. *Canadian Journal of Botany,* **53**, 87–93.

Lewin, R.A. (1956). Extracellular polysaccharides or green algae. *Canadian Journal of Microbiology,* **2**, 665–672.

Lewin, R.A. (1977). The use of algae as soil conditioners. *CIBCASIO Transactions (La Jolla, CA),* **3**, 31–35.

Lewin, R.A. (1983). Phycotechnology—How microbial geneticists may help. *BioScience,* **33**, 177–179.

Low, A.J. (1954). The study of soil structure in the field and in the laboratory. *Journal of Soil Science,* **5**, 57–74.

Low, A.J. (1962). The effects of organic matter on soil physical conditions especially structure. Welsh Soils Discussion Group, Report No. 3, 16–25.

Low, A.J. & Stuart, P.R. (1974). Micro-structural differences between arable and old grassland soils as shown by scanning electron microscopy. *Journal of Soil Science,* **25**, 135–137.

Lynch, J.M. (1977). Phytotoxicity of acetic acid produced in the anaerobic decomposition of wheat straw. *Journal of Applied Bacteriology*, **42**, 81–87.

Lynch, J.M. (1981). Promotion and inhibition of soil aggregate stabilisation by soil microorganisms. *Journal of General Microbiology*, **126**, 371–375.

Lynch, J.M. (1984). Interactions between biological processes, cultivation and soil structure. *Plant & Soil*, **76**, 307–318.

Lynch, J.M. & Elliot, L.F. (1983). Aggregate stabilisation of volcanic ash and soil during microbial degradation of straw. *Applied Environmental Microbiology*, **45**, 1398–1401.

Lynch, J.M., Ellis, F.B., Harper, S.H.T. & Christia, D.G. (1980). The effect of straw on the establishment and growth of winter cereals. *Agriculture & Environment*, **5**, 321–328.

Marshall, K.C. (1976). *Interfaces in Microbial Ecology*. Harvard University Press; Cambridge. Massachusetts.

Martin, J.P. (1942). The effect of composts and compost materials upon the aggregation of Collington sandy loam. *Soil Science Society of America Proceedings*, **7**, 218–222.

Martin, J.P. (1945). Microorganisms and soil aggregation. I. Origin and nature of some aggregating substances. *Soil Science*, **59**, 163–174.

Martin, J.P. (1971). Decomposition and binding action of polysaccharides in soil. *Soil Biology & Biochemistry*, **3**, 33–41.

Martin, J. & Anderson, D.A. (1943). Organic matter decomposition, mold flora and soil aggregation relationships. *Soil Science Society of American Proceedings*, **7**, 215–217.

Martin, J.P. & Richards, S.J. (1963). Decomposition and binding action of a polysaccharide from *Chromobacterium violaceum* in soil. *Journal of Bacteriology*, **85**, 1288–1294.

Martin, J.P. & Waksman, S.A. (1940). Influence of microorganisms on soil aggregation and erosion. *Soil Science*, **50**, 29–47.

Martin, J.P., Ervin, J.O. & Shepard, R.A. (1966). Decomposition of the iron, aluminium, zinc and copper salts or complexes of some microbial polysaccharides. *Soil Science Society of America Proceedings*, **30**, 196–200.

Martin, J.P., Parsa, A.A. & Haider, K. (1978). The influence of intimate association with humic polymers on biodegradation of [14C]-labelled organic substrates. *Soil Biology & Biochemistry*, **10**, 483–486.

McCalla, T.M. (1943). Influence of biological products on soil structure and infiltration. *Soil Science Society of America Proceedings*, **7**, 209–214.

McCalla, T.M. (1945). Influence of microorganisms and some organic substances on soil structure. *Soil Science*, **59**, 287–297.

McCalla, T.M. (1947). Influence of some microbial groups on stabilising soil structure against falling water drops. *Soil Science Society of America Proceedings*, **11**, 260–263.

McCalla, T.M. (1950). Microorganisms and soil structure. *Transactions of the Kansas Academy of Sciences*, **53**, 91–100.

McGill, W.B., Paul, E.A., Shields, J.A. & Lowe, W.E. (1973). Turnover of microbial populations and their metabolites in soil. *Bulletins from the Ecological Research Committee NFR* (Staten Naturvetensk Forshringstad, Stockholm), **17**, 292–301.

Mehta, N.C., Streuli, H., Muller, M. & Deuel, H. (1960). Role of polysaccharides in soil aggregation. *Journal of the Science of Food and Agriculture*, **11**, 40–47.

Menefee, E. & Hautala, E. (1978). Soil stabilisation by cellulose xanthate. *Nature (London)*, **275**, 530–532.

Metting, B. (1981). The systematics and ecology of soil algae. *Botanical Review*, **47**, 195–312.

Metting, B. & Rayburn, W.R. (1983). The influence of a microalgal conditioner on selected Washington soils: An empirical study. *Soil Science Society of America Journal*, **47**, 682–685.

Miller, D.E. & Kemper, W.D. (1962). Water stability of soils as influenced by incorporation of alfalfa. *Agronomy Journal*, **54**, 494–496.

Moavad, K., Bab'Yeva, I.P. & Gorin, S. Ye. (1976). Soil aggregation under the effect of extracellular polysaccharides of *Lipomyces eipofer*. *Soviet Soil Science*, **9**, 65–68.

Murray, R.S. & Quirk, J.P. (1985). Interparticle forces in relation to the stability of soil aggregates. In *Soil Colloids and Their Association in Soil Aggregates* (M.H.B. Hates & M.F.L. De Broodt, eds.). NATO Publications, Springer Verlag (in press).

Oades, J.M. (1978). Mucilages at the root surface. *Journal of Soil Science*, **29**, 1–16.

Oades, J.M. (1984). Soil organic matter and structural stability: organic mechanisms and implications for management. *Plant & Soil*, **76**, 319–337.

Oades, J.M. (1985). Natural associations of clays, hydrous oxides and organic matter. In *Soil Colloids and Their Associations in Soil Aggregates* (M.H.B. Hayes & M.F.L. De Boodt, eds.). NATO Publications, Springer Verlag (in press).

Olness, A. & Clapp, C.E. (1975). Influence of polysaccharide structure on dextran adsorption by montmorillonite. *Soil Biology & Biochemistry*, **7**, 113–118.

Peele, T.C. (1940). Microbial activity in relation to soil aggregation. *Journal of the American Society of Agronomy*, **32**, 204–212.

Reid, J.B. & Goss, M.J. (1981). Effect of living roots of different plant species on the aggregate stability of two arable soils. *Journal of Soil Science*, **32**, 521–541.

Rennie, D.A., Truog, E. & Allen, O.N. (1954). Soil aggregation as influenced by microbial gums, level of fertility and kind of crop. *Soil Science Society of America Proceedings*, **18**, 399–403.

Roychoudhury, P., Krishnamarti, G.S.R. & Venkataraman, G.S. (1980). Effect of algal inoculation on soil aggregation in rice soils. *Phykos.*, **19**, 224–227.

Sanatanatoglia, O.J. & Fernandez, M. (1983). Structural stability content of microbial gums under different types of management in a soil of the Ramallo series (vertic arguidol). *Ciencia de Suelo*, **1**, 43–49.

Sarkar, J.M. & Burns, R.G. (1984). Synthesis and properties of ß-D-glucosidase-phenolic copolymers as analogues of soil humic-enzyme complexes. *Soil Biology & Biochemistry*, **16**, 619–625.

Shanmuganathan, R.T. & Oades, J.M. (1982). Influence of anions on dispersion and physical properties of the A horizon of a red-brown earth. *Geoderma*, **29**, 257–277.

Stallings, J.H. (1953). Continuous organic matter supply—the key to soil aggregation and biological activity. *Journal of Soil and Water Conservation*. **8**, 178–184.

Stefanson, R.C. (1971). Effect of periodate and pyrophosphate on the seasonal changes in aggregate stabilisation. *Australian Journal of Soil Research*, **9**, 33–41.

Sutton, J.C. & Sheppard, B.R. (1976). Aggregation of sand dune soil by endomycorrhizal fungi. *Canadian Journal of Botany*, **54**, 326–333.

Swaby, R.J. (1949). The relationship between microorganisms and soil aggregation. *Journal of General Microbiology*, **3**, 236–254.

Swincer, G.D., Oades, J.M. & Greenland, D.J. (1968). Studies on soil polysaccharides: The composition and properties of polysaccharides under pasture and under a fallow-wheat rotation. *Australian Journal of Soil Research*, **6**, 225–235.

Theng, B.C.G. (1983). Clay polymer interactions: summary and perspectives. *Clays & Clay Minerals*, **30**, 1–10.

Tiedje, J.M., Sextone, A.J., Parkin, T.B., Revsbech, N.P. & Shelton, D.R. (1984). Anaerobic processes in soil. *Plant & Soil*, **76**, 197–212.

Tiessen, H.J. & Stewart, J.W.B. (1983). Particle size fractions and their use in studies of soil organic matter: II. Cultivation effects on organic matter composition in size fractions. *Soil Science Society of America Proceedings*, **47**, 509–514.

Tisdall, J.M. & Oades, J.M. (1979). Stabilisation of soil aggregates by the root system of ryegrass. *Australian Journal of Soil Research*, **17**, 429–441.

Tisdall, J.M. & Oades, J.M. (1980a). The effect of crop rotation on aggregation in a red-brown earth. *Australian Journal of Soil Research*, **18**, 423–433.

Tisdall, J.M. & Oades, J.M. (1980b). The management of ryegrass to stabilise the aggregates of a red-brown earth. *Australian Journal of Soil Research*, **18**, 412–422.

Tisdall, J.M. & Oades, J.M. (1982). Organic matter and water stable aggregates in soils. *Soil Science*, **33**, 141–163.

Tisdall, J.M., Cochcroft, B. & Uren, N.C. (1978). The stability of soils aggregates as affected by organic materials, microbial activity and physical disruption. *Australian Journal of Soil Research*, **16**, 9–17.

Turchenek, L.W. & Oades, J.M. (1978). Organo-mineral particles in soils. In *Modification of Soil Structure* (W.W. Emerson, R.D. Bond & A.R. Dexter, eds.). pp. 137–144. John Wiley & Sons; Chichester.

Turchenek, L.W. & Oades, J.M. (1979). Fractionation of organo-mineral complexes by sedimentation and density techniques. *Geoderma*, **21**, 311–343.

Verma, L. & Martin, J.P. (1976). Decomposition of algal cells and components and their stabilisation through complexing with model humic acid-type phenolic polymers. *Soil Biology*

& *Biochemistry*, **8**, 85–90.
Webber, L.R. (1965). Soil polysaccharides and aggregation in crop sequences. *Soil Science Society of America Proceedings*, **29**, 39–42.

Recycling of Organic Wastes for a Sustainable Agriculture

J.F. Parr, R.I. Papendick and D. Colacicco

U.S. Department of Agriculture, Beltsville, Maryland, U.S.D.A., Pullman, Washington; and U.S.D.A., Washington, D.C., U.S.A., respectively[1]

INTRODUCTION

A growing number of agricultural scientists, environmentalists, government officials, farmers, and both urban and rural laymen, have become increasingly alarmed over the potential vulnerability of the energy-intensive systems of food and fiber production which now characterize U.S. agriculture. During the past 40 years, conventional agriculture has become increasingly dependent upon petroleum-based, chemically-synthesized fertilizers and pesticides for crop protection and to supply plant nutrients. Certainly, these energy-intensive technologies have contributed greatly to this Nation's agricultural productivity. However, sharply escalating production costs associated with the increasing cost and uncertain availability of energy, i.e. fuel and fertilizers, have generated considerable interest in less expensive and more environmentally compatible production alternatives such as organic farming (USDA, 1980).

At a recent multi-agency workshop involving more than 100 prominent scientists and administrators, "Sustaining the Soil Productivity"[2] was unanimously selected as our foremost national agricultural research priority (Larson *et al.,* 1981). Today in the U.S. Corn Belt, which contains much of country's prime farmland, and where intensive row-cropping is practiced, the average annual soil loss from erosion exceeds 8 tons/acre (18 mt/ha) (Berg, 1979). This is about twice the maximum tolerable rate or so-called "T-value"

---

[1]Soil Microbiologist and Soil Scientist, Agricultural Research Service, and Agricultural Economist, Economic Research Service

[2]Soil Productivity as defined in "Soil", the 1957 USDA Yearbook of Agriculture, is "The capability of a soil for producing a specified plant or sequence of plants under a defined set of management practices. It is measured in terms of outputs or harvests in relation to the inputs of production factors for a specific kind of soil under a physically defined system of management."

that will sustain a reasonably high level of soil productivity. The apparent decline in soil productivity throughout the U.S. from excessive soil erosion, nutrient runoff, and loss of soil organic matter; the impairment of environmental quality from sedimentation and pollution of natural waters by agricultural chemicals; and the potential hazards to human and animal health and food safety from heavy use of pesticides, have also stimulated interest in organic farming systems of food production.

The purpose of this paper is to present some new perspectives and strategies for efficient and effective use of organic wastes to enhance sustainable systems of agriculture in both developed and developing countries.

TRADITIONAL USE OF ORGANIC WASTES AND RESIDUES IN AGRICULTURE

Most countries have traditionally utilized various kinds of organic materials to maintain or improve the tilth, fertility, and productivity of their agricultural soils. However, several decades ago organic recycling practices in some countries were largely replaced with chemical fertilizers which were applied to high yielding cereal grains that responded best to a high level of fertility and adequate moisture, including irrigation. Soil cultivation was also intensified to improve weed control and seedbed conditions. Consequently, the importance of organic matter to crop production received less emphasis, and its proper use in soil management was neglected, or even forgotten. As a result of this, and failure to implement effective soil conservation practices, the agricultural soils in a number of developed and developing countries have undergone serious degradation and decline in productivity because of excessive soil erosion and nutrient runoff, and the decrease in stable soil organic matter levels.

The most realistic approach for many of the developing countries to achieve sufficiency in food production and maximum crop yield potential is to accelerate efforts to halt the decline in soil productivity and to restore the productivity of degraded soils in the shortest possible time. A recent FAO assessment of organic recycling states that the improvement of soil productivity as a whole is expected to contribute about 60 percent of the increased food production that is currently needed worldwide (Hauck, 1981). Much of this goal can be achieved through proper management of agricultural and municipal organic wastes on land to protect agricultural soils from wind and water erosion and to prevent nutrient losses through runoff and leaching. Efficient and effective use of these materials as soil conditioners also provides one of the best means we have for maintaining and restoring soil productivity. The beneficial effects of organic wastes on soil physical properties as evidenced by increased water infiltration, water-holding capacity, water content, aeration and permeability, soil aggregation and rooting depth, and by decreased soil crusting, bulk density, and runoff and erosion are widely known (USDA, 1957).

SURVEYS OF ORGANIC MATERIALS

Most of the countries lack reliable information as to the types, amounts, and availability of different organic wastes that might be utilized to improve the productivity of their agricultural soils. Such information is highly essential as a first step for successful planning and implementation of organic recycling programs. One such survey that could serve as a model is a recent report entitled "Improving Soils with Organic Wastes" submitted by the U.S. Department of Agriculture to the Congress on "the practicability, desirability, and feasibility of collecting, transporting, and placing organic wastes on land to improve soil tilth and fertility." This information was urgently needed because of the steadily increasing costs of energy, fertilizers, and pesticides to U.S. farmers and the problems of soil deterioration and erosion associated with intensive farming systems (USDA, 1978). The report contains detailed information on the availability of seven major organic waste materials for use in improving soil tilth and fertility, i.e., (a) animal manures, (b) crop residues, (c) sewage sludge, (d) food processing wastes, (e) industrial organic wastes, (f) logging and wood manufacturing wastes, and (g) municipal refuse. Information is reported on the quantity currently generated, present usage, potential value as fertilizers, competitive uses, and problems and constraints affecting their use.

A summary of the USDA report is presented in Table 1. A total of about 730 million dry metric tons of organic wastes are produced annually. This represents a national resource of significant economic value, and its proper and efficient use should be emphasized. About 50% of this total is comprised of crop residues, while about 22% is made up of animal manures. Thus, nearly three quarters of the total annual production of organic wastes in the U.S. is associated with crop residues and animal manures. The USDA report established that about 75% of these two wastes are currently being applied to land for improving soil productivity.

Once these surveys have been made countries within a region should exchange them for mutual interest and benefit. For example, wastes from several processing operations might be co-composted to produce higher quality organic amendments for soil improvement and plant growth. Data in these surveys should be kept current by updating at 2- to 3-year intervals because as cities and industries continue to grow, waste production will also increase. Both the United Nations Environment Programme (UNEP) and the Food and Agriculture Organization (FAO) have emphasized the need for basic information on waste generation and utilization in developing countries (FAO, 1977).

TABLE 1

Annual production of organic wastes in the United States,
current use on land, and probability of increased use

| | Total Production | | | |
Organic wastes	Dry metric tons (×1000)	% of total	Current use on land[z] (%)	Probability of increased use[y] on land
Animal manure	158,730	21.8	90	Low
Crop residues	391,009	53.7	68	Low
Sewage sludge and septage	3,963	0.5	23	Medium
Food processing	2,902	0.4	(13)	Low
Industrial organic	7,452	1.0	3	Low
Logging and wood manufacturing	32,394	4.5	(5)	Very low
Municipal refuse	131,519	18.1	(1)	Low
TOTAL	727,969	100.0	—	—

[z]Values in parentheses are estimates because of insufficient data.
[y]Medium indicates a likely increase of 20 to 50%, low indicates a 5 to 20% increase, and very low indicates less than 5% increase.

Recently, FAO published a world directory of institutions concerned with the utilization of agricultural residues (FAO, 1978a); a compendium of technologies for treatment of agricultural residues (FAO, 1978b); and a bibliography of papers and reports dealing with the utilization of agricultural residues (FAO, 1978c). More recently, FAO published a survey on the utilization of residues from agriculture, forestry, fisheries, and related industries which included responses from 57 countries (FAO, 1979). However, only a few countries were able to provide comprehensive information on the kinds and amounts of organic wastes or residues being generated, or their availability for utilization as soil amendments and organic fertilizers.

CONSTRAINTS AND POTENTIAL FOR INCREASED USE OF ORGANIC WASTES

Sewage sludge makes up about 0.5% of the total organic waste produced in the United States and approximately a quarter of it is currently applied to land. The other four wastes listed in Table 1 have not been used extensively on land because of certain competitive uses, high costs of collection, processing, transportation, and application; and because of constraints on usage related to certain chemical and physical properties. For example, (a) cotton gin trash and sugarcane bagasse are now increasingly sought as sources of fuel for burning, (b) some food processing wastes may have extremely high acidity or alkalinity that may adversely affect soil pH, (c) some sewage sludges contain excessive amounts of heavy metals and organic chemicals that may be toxic to plants or endanger the food chain after absorption and accumulation, and (d) shredded municipal refuse may contain considerable amounts of solid fragments (glass, plastic, and metal) that do not readily biodegrade and might detract aesthetically when applied to land.

The potential for increased use of organic wastes on land to improve the productivity of soils in the USA is low (Table 1). Only the use of municipal sewage sludge on land is expected to increase appreciably, but this increase is very small when compared on a national basis with the two largest waste categories, i.e., animal manures and crop residues.

The USDA (1978) report pointed out that there is a growing shortage of good quality organic wastes for use in maintaining and improving the productivity of our agricultural soils. The report cited a number of ways in which our limited amounts of organic wastes might be used more effectively as soil amendments. These include:

(1) Improving methods of collection, storage, and processing (e.g., composting) of animal manures to minimize the loss of nitrogen that often occurs in these operations.
(2) Applying manures to land that are presently being wasted.
(3) Applying crop residues to land that are not now being fully utilized.
(4) Increasing the use of sewage sludge on land.
(5) Increasing the use of the organic/compostable fraction of municipal refuse.

REINTRODUCTION OF ORGANIC FERTILISERS

Organic fertilizers, including animal manures, crop residues, green manures, and composts were traditionally and preferentially used in developing countries until the 1960s when chemical fertilizers began to gain in popularity. Chemical fertilizers became easily available and unlike organic fertilizers they were less bulky and, thus, easier to transport, handle, and store. They were also relatively inexpensive and produced more striking results than organic fertilizers, particularly during the era of the "Green Revolution" when crop varieties were introduced that responded best to heavy applications of chemical fertilizers. Thus, when the world energy crisis began in the early 1970s, chemical fertilizers had virtually replaced organic sources of crop nutrients in developing countries (FAO, 1975). Because of steadily rising energy costs, chemical fertilizers have become much more expensive than they once were, and now organic fertilizers have started to regain their lost popularity.

A significant consequence of these events, as Hauck (1978) points out, is that "in many countries that have recently been increasingly dependent upon mineral fertilizers, the technical knowledge of organic waste utilization has been lost. It is thus necessary to reintroduce the established techniques, to improve them, and to develop new practices conforming to modern technology."

The shift away from organic recycling practices also served to re-emphasize the value of, and need for, organic amendments for the short- and long-term improvement of cultivated soils and maintenance of soil productivity. Without regular additions of adequate amounts of organic materials to soils, there is increased leaching, erosion, and gradual deterioration of their physical properties. Moreover, as the soil degrades, there is a concomitant decrease in the crop use efficiency of chemical fertilizers, especially nitrogen.

Evironmental pollution has also become an international concern. Thus, proper processing and recycling of organic wastes as resources for agriculture can greatly reduce environmental pollution. Additional benefits include improved public health, conservation of resources, and better appearance of both urban and rural communities.

REINTRODUCTION OF BEST MANAGEMENT PRACTICES

Many farmers in developing countries have shifted toward intensive row cropping for short-term economic gain, and have thus neglected soil and water conservation practices. This has resulted in markedly increased soil erosion, extensive damage to cropland, sedimentation, and nutrient runoff and enrichment of surface waters. With increasing public concern for environmental pollution and agriculture's contribution to it, there is an urgent need to reintroduce those soil and crop management practices which have been cited as best management practices for controlling soil erosion and water pollution from cropland (USDA/EPA, 1975, 1976). These include the use of sod-based rotations, contouring, conservation tillage, cover crops, grassed waterways, and possibly others such as divided slope farming. Research and extension programs should be formulated and implemented, in both developed and developing countries, to demonstrate the cost/benefit relationships of conservation management practices. Aspects of multiple cropping systems such as double cropping, sequential cropping, and intercropping may also provide special means for controlling wind and water erosion, and for effective recycling of nutrients from crop residues (ASA, 1976).

COMPOSTING TO ENHANCE THE USEFULNESS AND ACCEPTABILITY OF ORGANIC WASTES

One way in which some of the problems associated with the utilization of various organic wastes (e.g. odors, human pathogens, and storage and handling constraints) can be resolved is by composting. Composting is an ancient practice whereby farmers have converted organic wastes into resources that provide nutrients to crops and enhance the tilth, fertility, and productivity of soils. Through composting, organic wastes are decomposed, nutrients are made available to plants, pathogens are destroyed, and malodors are abated. The historical aspects of composting have been thoroughly discussed in reviews by Gotaas (1956) and Golueke (1972).

Recently, the U.S. Department of Agriculture at Beltsville, Maryland, developed the highly successful Beltsville Aerated Rapid Composting (BARC) Method for composting sewage sludge, animal manures, municipal refuse, and pit latrine wastes (Willson *et al.*, 1980). This method has been widely adopted by both large and small municipalities throughout the United States for composting sewage sludge and solid waste. A number of developing countries have also adopted this technology for composting. The method is simple and relatively inexpensive, yet effective, and allows considerable trade-off between labor and capital.

Composts provide a more stabilized form of organic matter than raw wastes and can vastly improve the physical properties of soils. For example, addition of sludge compost to sandy soils will increase their ability to retain water and render them less droughty. In heavy-textured clay soils, the added organic matter will increase permeability to air, and increase water infiltration thereby minimizing surface runoff and increasing water storage. Addition of sludge compost to clay soils has been shown to reduce soil compaction, lower the bulk density, and increase the rooting depth.

Recent reports by Hornick *et al.* (1979, 1984) discusses the uses of sewage sludge compost for soil improvement and plant growth including (a) establishment, maintenance, and production of turfgrass and sod, (b) use in vegetable gardens, (c) production of field crops and forage grasses, (d) use on nursery crops and ornamentals, (e) use in potting mixes, and (f) reclamation and revegetation of disturbed lands. Recommendations are provided as to time, methods, and rates of compost application for different soils and management practices.

Some organic wastes may have chemical, physical, and/or microbiological properties that would greatly limit the extent to which they could be composted alone. For example, some wastes may have an extremely acidic or alkaline pH, others may have an unusually high or low C:N ratio, and still others may vary widely in their solids content. In such cases, selective co-composting of these wastes with sewage sludges, pit latrine waste or night soil, municipal solid waste (i.e., garbage or refuse), crop residue, animal manures, food processing wastes and certain industrial wastes, may alleviate these deficiencies and provide a readily compostable mixture and higher quality product.

CONSIDERATION OF ORGANIC METHODS OF FARMING

According to a recent report on organic farming by the U.S. Department of Agriculture (USDA, 1980), there are several thousand farmers in the U.S. operating large- and small-scale commercial farms profitably with minimal or no use of chemical fertilizers or pesticides. They are referred to as organic farmers and, although there is considerable diversity in their individual farming methods, they collectively advocate that sustainable and successful farming systems are based primarily on the proper care and protection of the soil.

Organic farming employs basically a systems approach to farm management. On well-managed farms the various practices used are often interrelated so that each augments the other to form a complex but efficient production system. For example, a legume may be grown in a crop rotation not only to produce feed for animals but to control specific weeds and insects, and to supply nitrogen for grain crops that follow. Most organic farmers in the U.S. rely heavily on recycling of organic materials and use of green manure crops and legumes in the rotation to supply nutrients, and to maintain nutrient balances and soil organic matter. As a group, organic farmers are highly committed to protecting the soil resource which they do by regular use of sod-based rotations, legumes, animal manures, and other organic wastes. The USDA Report revealed that organic farming methods and practices were effective in controlling soil erosion and nutrient runoff and in minimizing environmental pollution. Organic farming is also practiced to a limited extent in other developed countries as well, including Japan and those of Western Europe.

The information in the USDA Report (1980) is highly relevant to a number of agricultural problems and concerns that exist today in both developed and developing countries, such as environmental pollution, the high cost of energy, the high cost and uncertain availability of chemical fertilizers, the lack of effective soil and water conservation practices, and the need for improving public health and food safety and quality. There is an urgent need to develop long-term, low-energy, biological, self-sustainable systems of farming in developing countries and developed countries as well. The extent to which some of the organic methods and practices used in more developed countries can be adapted beneficially in meeting this goal should be thoroughly explored.

SPECIAL MANAGEMENT PRACTICES FOR UTILIZATION OF ORGANIC WASTES ON LAND

Special management practices may be needed if we are to obtain the full value of organic wastes as soil amendments and as sources of plant nutrients. For example, the value of organic materials as fertilizers is increased if nutrient release through decomposition and mineralization coincides with the crop's nutrient requirement curve. The nutrient release pattern of organic materials can be controlled to some extent by proper timing of application, pretreatment methods such as composting, and by the method of application. For example, certain organic materials may decompose more slowly if left on the soil surface or concentrated in the soil than if thoroughly mixed in the tillage layer (Parr & Papendick, 1978).

There are many problems involved in handling and applying organic wastes to land because their physical characteristics are so variable. Such wastes are almost always bulky, and can range from liquids on one extreme to dry solid material on the other. Existing application technologies are often ineffective and fail to achieve the desired level of erosion control or increased crop production. Methods of application must be simple, inexpensive, energy conserving, and effective for nutrient recycling and erosion control.

An example of a particularly effective soil erosion control measure for sloping lands is that of vertical mulching, a procedure which incorporates organic materials such as crop residues into a vertical channel 30 to 40 cm deep and some 10 to 12 cm wide at the soil surface (Parr, 1959). The operation is usually conducted on the contour with intervals ranging from 5 to 10 m. A recent modification of this procedure is referred to as "slot mulch" and has been demonstrated as an effective means of controlling erosion on steeply sloping lands in the winter wheat area of eastern Washington State (Saxton *et al.,* 1981). Both techniques are highly effective in intercepting runoff, enhancing root penetration and development, and providing for effective water conservation and storage for crop use. Although several types of machines have been designed for these operations, similar results could be achieved using hand labor.

This concept of residue management may have application in developing countries for restoring the productivity of sloping lands that have suffered from severe soil erosion. Besides crop residues, other types of organic materials, including composts, could also be utilized in the vertical mulch or slot mulch procedures.

NON-TRADITIONAL SOIL AND PLANT ADDITIVES

There are a number of products that have been introduced into developing countries, that are generally referred to as soil and plant additives, for which the manufacturers' claims greatly exceed the performance of the product (Weaver *et al.*, 1974; Dunigan, 1979; Weaver, 1979).

These products include (a) *microbial fertilizers* and *soil inoculants* which are purported to contain unique and beneficial strains of soil micoorganisms, (b) *microbial acitvators* that supposedly contain special chemical formulations for increasing the numbers and activity of beneficial microorganisms in soil, (c) *soil conditioners* that claim to create favorable soil physical and chemical conditions which result in increased growth and yield of crops, and (d) *plant stimulants and growth regulators* that supposedly stimulate plant growth, resulting in healthier and more vigorous plants, and increased yields.

In most cases where researchers have evaluated these products using acceptable scientific and statistical methods they have been unable to demonstrate any significant yield increases. Such studies have also usually failed to provide evidence for any additional claims of benefit. It is noteworthy, however, that there are some legitimate products on the market that have stood the test of time. A classic example is the commercial preparation of the nitrogen fixing bacteria *Rhizobium* used for inoculating legume seeds.

The proper use of organic fertilizers, chemical fertilizers, lime, and specific rhizobia inoculum for legumes, when needed, will usually pay dividends to farmers. However, any product which promises to perform extraordinary processes in soils and plants, or to have magical and mysterious beneficial effects on plants and microorganisms when applied at very low rates of application should be viewed with caution and scepticism. Such products are invariably a poor investment, of little or no economic value, and cannot substitute for good farming methods and sound management practices in either developed or developing countries.

VALUE OF ORGANIC MATERIALS AS FERTILIZERS AND SOIL CONDITIONERS

The "value" of organic materials as fertilizers and soil conditioners is often misunderstood and has been the source of some controversy. The simplest and most common means of estimating the value of organic amendments is by assessing the current market value of the plant nutrients they contain. Usually this is done in terms of their macronutrient content, i.e., nitrogen, phosphorus, and potassium.

However, many organic materials contain other components which can contribute significantly to increased crop yields, including organic matter, secondary and micronutrients, and sometimes lime. In some cases, the organic matter fraction of a particular material may have a higher value than that of its total nutrient content because of the beneficial effect of organic matter on soil physical properties and improvement of soil productivity. A brief clarification of the agronomic and economic values of organic materials follows.

The Agronomic Value

The agronomic value of an organic material is the increased crop yield or quality delivery from its application. There is a considerable amount of data in the literature which demonstrates the effect of organic materials on crop yield, but very little on crop quality. Crop yield response to additions of organic materials is highly variable and is dependent upon the crop, soil type, climatic conditions, management system, and the organic material used. In most cases, crop yield response to the addition of organic materials is non-linear. The greatest yield response is obtained with the first few increments of organic material, followed by progressively smaller yield increases with additional increments. Thus, as with chemical fertilizers and other production inputs, crop yield response to an organic amendment follows the law of diminishing returns, and is responsible for the decreased value per unit of the material with increased application rates. Obviously, both the agronomic and economic value per unit of organic material to the farmer are correspondingly higher at low, rather than high, application rates. Reliable estimates of the economic value of organic materials depends upon the accurate assessment of agronomic data relating the crop yield response to the application of a particular material.

The Economic Value

The economic value of an organic waste or residue to a farmer is the value of the increase in crop yield and/or crop quality that is derived from its use. Since the crop yield response to an organic amendment follows the law of diminishing returns, the average yield increase is always larger than the yield increase attributable to the incremental unit of organic material. If the price of the farmer's product is the same regardless of the quantity produced, which is usually the case, then the revenue derived from the average unit of organic material will be greater than the revenue from the incremental unit. Thus the value of the average unit of organic material will be higher than the value of the incremental unit of organic material.

Farmers can be expected to utilize organic materials to the point where the revenue from the incremental unit is equal to its price, assuming application is included in the price. They would certainly use no more than that quantity since additional application of the organic material would yield a loss.

Likewise, they would not wish to use less since total profits would decline. Thus an estimate of the value of the incremental unit of organic material would be useful to farmers as well as public waste management agencies analyzing the marketability of a recycled product. However, public decision-makers should also be interested in estimates of the total economic benefits from resource recycling alternatives, such as composting, to compare them to destructive alternatives such as landfilling or incineration. Thus, an estimate of the value of the average unit of organic material would be the appropriate figure to use in public project analysis.

The rate at which organic materials decompose or mineralize in soil is highly variable but they do have a greater residual effect on soil fertility than most chemical fertilizers because of the slow-release character of the nitrogen and phosphorus components. Thus, a significant portion of the value of organic materials as fertilizers is their capacity to elicit yield responses from succeeding crops. This response must be accounted for to assess the true value of the material. Barbarika *et al.* (1980) estimated that the cumulative economic value of some organic materials applied to agricultural soils could be as much as five times greater in succeeding years than the value realized during the application year.

RESEARCH NEEDS

In addition to traditional organic materials such as animal manures and crop residues, developing countries are increasing their land application of municipal wastes (e.g., sewage sludges and effluents, and garbage) and industrial wastes (e.g., food processing and acceptable industrial organic wastes). Research is needed to determine how these wastes differ in their ability to improve the tilth, fertility, and productivity of soils.

Information is limited on the substitutability of one particular organic waste for another in soil improvement. Criteria should be developed by which the relative effectiveness of different organic wastes can be compared. For example, studies are needed to determine the following properties: decomposition rates for each waste under different soil regimes and cropping systems; rates at which plant nutrients are mineralized, recycled, and utilized by both current and subsequent crops; the potential toxic effects of certain wastes on plants and microorganisms; the impact of organic waste management on the control of plant insects and diseases; and the extent to which different wastes can effect desirable and residual improvement of soil physical properties. Each organic waste has unique properties that should be thoroughly investigated in the soil/water/plant ecosystem.

Some organic amendments are known to mineralize and release available plant nutrients rapidly as a result of microbial attack. In some cases this is desirable, particularly on soils that are already in a high state of fertility and productivity. On the other hand, marginal, erodible, sloping, and generally less productive soils would benefit, at least initially, from application of organic materials having a higher degree of microbial stability in soil. Such materials would release their plant nutrients at a relatively slower rate. Farmers in developing countries often have occasion to use both types of materials in their farming operations depending on whether there is a need to release nutrients rapidly, or to improve the productivity of marginal soils. Table 2 lists some organic materials that would be expected to differ considerably in these two properties. Although these are hypothetical values and would have to be verified experimentally, the concept introduced here is the important consideration. A high nutrient availability index (NAI) indicates materials that would release nutrients relatively rapidly, while a high organic stability index (OSI) would be associated with more stable forms of organic matter. Materials with a high NAI value usually would be expected to have a low OSI value, and conversely. Research is needed to develop reliable numerical indexes such as those shown in Table 2. This would allow a realistic basis for predicting the nutrient availability and organic stability of different wastes under different soil, climatic, and cropping conditions.

As noted earlier co-composting can overcome certain chemical, physical or microbiological deficiencies of some organic wastes. For example, wastes such as rice straw that have very high C:N ratios, and which would compost slowly alone, might be combined with a low C:N material such as poultry manure to achieve a more favorable ratio for composting. Research is needed to determine the proper combination of different organic wastes that could be

TABLE 2

Hypothetical nutrient availability indexes (NAI) and organic stability indexes (OSI) for composted and uncomposted organic materials

Product	Nutrient Availability Index[a]	Organic Stability Index[a]
Beef Manure	70	30
Rice Straw	15	85
Sewage Sludge	80	20
Sewage Sludge – woodchip compost	35	65
Refuse compost	25	75
Refuse-pit latrine compost	40	60
Poultry manure – rice straw compost	65	35

[a]Values are hypothetical, but are based mainly on what is known of the C:N ratio of organic materials cited, and the decomposability of the various components (i.e. cellulose, hemicellulose, lignin, sugars, proteins, etc.) that comprise them (Parr & Papendick, 1978).

successfully co-composted. The effectiveness of these composts for improving soil productivity should be throughly explored.

It is unlikely that organic fertilizers will totally replace chemical fertilizers in developing countries, nor should that be the goal. There is evidence that higher crop yields are possible when organic wastes are applied in combination with chemical fertilizers than when either one is supplied alone, so that organic amendments may increase the efficiency of chemical fertilizers. Research is needed to evaluate the effect of various combinations of organic amendments and chemical fertilizers on crop yields and fertilizer efficiency. The potential for enriching (i.e., spiking) organic wastes and composts with chemical fertilizers or other wastes of a higher plant nutrient content to enhance their fertilizer value should be thoroughly evaluated.

Farmers frequently apply organic wastes (including composts) to their fields at rates that are too low or too high for maximum economic return. When the rate is too low, soil physical properties are not sufficiently improved and the plant nutrient level is inadequate to sustain optimum crop growth and yield. If applied at excessive rates, plant nutrients are not utilized efficiently and contribute to environmental pollution through runoff and leaching. Research is needed to improve the efficient and effective use of organic materials in cropping systems. Such research must consider a number of management factors and how these materials are applied, including: mode or method of application (i.e., surface-applied, plowed-down, disked in or side-dressed); rate, time, and frequency of application; the soil, and the sequence of crops to be grown. A better understanding of the interaction of these factors should provide a more reliable basis for developing more efficient and effective waste application/utilization programs for agriculture.

The beneficial effects of various organic materials on soil productivity are well known. However, for most organic materials it is very difficult to quantify what portion of the crop yield response is due to the organic matter fraction and what is due to the plant nutrient content. Experimental procedures and special data analysis techniques should be thoroughly explored so that the economic value of the organic component in different kinds of organic materials can be accurately and meaningfully estimated. This should be done for a number of different crops, soils, management systems, and climatic situations. Such data could greatly stimulate the use of organic wastes on agricultural land to control soil erosion and nutrient runoff and to improve soil productivity.

References

American Society of Agronomy (1976). *Multiple Cropping*. ASA Special Publication No. 27. Madison; Wisconsin. 378 pp.

Barbarika, A., Colacicco, D. & Bellows, W.J. (1980). The value and use of organic wastes. Maryland Agri-Economics, May 1980. Cooperative Extension Service, University of Maryland; College Park, Maryland. 5 pp.

Berg, N.A. (1979). Soil conservation; the physical resource setting. In *Soil Conservation Policies: An Assessment,* pp. 8–17. Soil Conservation Society of America; Ankeny, Indiana.

Dunigan, E.P. (1979). Microbial fertilizers, activators and conditioners: a critical review. *Developments in Industrial Microbiology,* **20**, 311–322.

F.A.O. (1975). *Organic materials as fertilizers.* Soils Bulletin No. 27. Food and Agriculture Organisation, United Nations; Rome. 394 pp.

F.A.O. (1977). *Residue Utilization–Management of Agricultural and Agro-Industrial Wastes.* Report of UNEP/FAO Seminar held in Rome, January 18–21, 1977. 67 pp.

F.A.O. (1978a). *Agricultural residues: world directory of institutions.* Agricultural Services Bulletin No. 21. 2nd. edition. Food and Agriculture Organisation, United Nations; Rome 30 pp.

F.A.O. (1978b). *Agricultural residues: compendium of technologies.* Agricultural Services Bulletin No. 33. Food and Agriculture Organisation, United Nations; Rome. 370 pp.

F.A.O. (1978c). *Bibliography of agricultural residues.* Agricultural Services Bulletin No. 35. Food and Agriculture Organisation, United Nations; Rome. 125 pp.

F.A.O. (1979). *Agricultural residues: quantitative survey.* Agricultural Services Bulletin. Food and Agriculture Organisation, United Nations; Rome. 116 pp.

Golueke, C.G. (1972). *Composting: A Study of the Process and its Principles.* Rodale Press; Emmaus, PA. 110 pp.

Gotaas, H.B. (1956). *Composting. Sanitary Disposal and Reclamation of Organic Wastes.* World Health Organisation Monograph 31. Geneva; Switzerland. 205 pp.

Hauck, F.W. (1978). Organic recycling to improve soil productivity. Paper presented at the FAO/SIDA Workshop on "Organic Materials and Soil Productivity in the Near East". Food and Agriculture Organisation, United Nations; Rome.

Hauck, F.W. (1981). The importance of organic recycling in the agricultural programmes of developing countries. In *Organic Recycling in Asia,* pp. 17–28. FAO/UNDP Regional Project RAS/75/004. Project Field Document No. 20. Food and Agriculture Organisation, United Nations; Rome. 184 pp.

Hornick, S.B., Murray, J.J., Chaney, R.L., Sikora, L.J., Parr, J.F., Burge, W.D., Willson, G.B. & Tester, C.F. (1979). Use of sewage sludge compost for soil improvement and plant growth. Science and Education Administration, Agricultural Reviews and Manuals, Northeastern Region Series 6. U.S. Department of Agriculture; Beltsville, MD. 10pp.

Hornick, S.B., Sikora, S.B., Sterrett, S.B., Murray, J.J., Millner, P.D., Burge, W.D., Colacicco, D., Parr, J.F., Chaney, R.L. & Willson, G.B. (1984). Utilization of sewage sludge compost as a soil conditioner and fertilizer for plant growth. Agriculture Information Bulletin No. 464. U.S. Department of Agriculture; Beltsville, MD. 32pp.

Larson, W.E., Walsh, L.M., Stewart, B.A. & Boelter, D.H. eds. (1981). *Soil and water resources: research priorities for the nation (executive summary).* Soil Science Society of America, Inc.; Madison, WI. 45 pp.

Parr, J.F. (1959). Effects of vertical mulching and subsoiling on soil physical properties. *Agronomy Journal,* **51**, 412–414.

Parr, J.F. & Papendick, R.I. (1978). Factors affecting the decomposition of crop residues by microorganisms. In *Crop Residue Management Systems,* pp. 101–129. American Society of Agronomy; Madison, WI. 248 pp.

Saxton, K.E., McCool, D.K. & Papendick, R.I. (1981). Slot mulch for runoff and erosion control. *Journal of Soil & Water Conservation,* **36**, 44–47.

U.S. Department of Agriculture (1957). *Soil–Yearbook of Agriculture.* U.S. Government Printing Office; Washington, D.C.

U.S. Department of Agriculture/U.S. Environmental Protection Agency (1975) Joint Publication on *"Control of Water Pollution from Cropland".* Volume I: A Manual for Guideline Development. U.S. Government Printing Office; Washington, D.C. 111 pp.

U.S. Department of Agriculture/U.S. Environmental Protection Agency (1976) Joint Publication on *"Control of Water Pollution from Cropland".* Volume II. An Overview. U.S. Government Printing Office; Washington, D.C. 187 pp.

U.S. Department of Agriculture (1978). *Improving Soils with Organic Wastes.* Report to the

Congress in response to Section 1461 of the Food and Agriculture Act of 1977 (PL 95-113). U.S. Government Printing Office; Washington, D.C. 157 pp.

U.S. Department of Agriculture (1980). *Report and Recommendations on Organic Farming*. A special report prepared for the Secretary of Agriculture. U.S. Government Printing Office; Washington, D.C. 94 pp.

Weaver, R.W. (1979). Evaluation of the effectiveness of microbial fertilizers, activators and conditioners. *Developments in Industrial Microbiology*, **20**, 323–327.

Weaver, R.W., Dunigan, E.P., Parr, J.F. & Hiltbold, A.E. (1974). Effect of two soil activators on crop yields and activities of soil microorganisms in the Southern United States. Southern Cooperative Series Bulletin No. 189. 24 pp.

Willson, G.B., Parr, J.F., Epstein, E., Marsh, P.B., Chaney, R.L., Colacicco, D., Burge, W.D., Sikora, L.J., Tester, C.F. & Hornick, S.B. (1980). *Manual for composting sewage sludge by the Beltsville aerated pile method*. Joint publication by U.S. Department of Agriculture/U.S. Environmental Protection Agency. EPA-600/8-80-022. U.S. Government Printing Office; Washington, D.C. 64 pp.

Crop Residue Management for Improved Soil Productivity[1]

L.F. Elliott and R.I. Papendick

U.S. Department of Agriculture, Agricultural Research Service, Pullman, Washington, 99165, U.S.A.

INTRODUCTION

Frequently, crop residues are viewed as a nuisance and a disposal problem. This is particularly true when crops are seeded into heavy surface or shallow-incorporated residues, which can result in large crop yield reductions (Lovett *et al.*, 1982; Lynch & Elliott, 1984). Yield reductions have also been documented when large quantities of residues are completely incorporated into the soil (Cannell *et al.*, 1982). Poor stands, crop growth and yields are attributed to the residue problem for a variety of reasons including the production of compounds toxic to crop growth during residue decomposition, favourable conditions for the development of pathogen problems, root surface colonists that impede plant growth, and poor nutrient relationships. For these reasons, burning and complete incorporation of residues are widely accepted disposal practices. However, there are negative aspects associated with residue disposal. Burning the residues destroys much of the nitrogen contained in the residues, may result in severe soil erosion problems, destroys any soil conditioning effects of the residues, modifies the ecological effects of the residue on the soil microflora, and causes the soil surface to be hydrophobic, which can reduce infiltration and water storage (Lynch & Elliott, 1984). Tillage, especially intensive tillage, can increase soil erosion, result in the loss of valuable soil organic matter, and may be detrimental to soil structure (Low, 1972; Lovett *et al.*, 1982; Stevenson, 1982; Tisdall & Oades, 1982). Burying of low nitrogen-containing crop residues also likely prevents

[1]Contribution from the Agricultural Research Service, U.S. Department of Agriculture, in cooperation with the College of Agriculture Research Center, Washington State University, Pullman, WA., 99164, U.S.A. W.S.U. Scientific Paper No. 7202.

maximum benefits of the residues to soil structure at the soil surface as compared with surface-managed residues. It is our intent to discuss crop residue decomposition problems associated with crop residues and possibilities for residue management for maximum utilization of this resource.

THE INFLUENCE OF CROP RESIDUE DECOMPOSITION ON SOIL AGGREGATION AND SOIL ORGANIC MATTER

Soil aggregation

It is well established that microorganisms are beneficial to soil structure and that soil aggregation can be increased by the addition of crop residues presumably due to the resultant soil microbial activity during the decomposition process (Gilmour *et al.*, 1948; Lovett *et al.*, 1982; Martin, 1971; Stevenson, 1982). We know that microbial cells and their excretory products can increase soil aggregation (Lynch, 1981, 1984); however, each might serve different roles depending on the environment, how the residues are managed, and chemical composition of the residues. For example, during the decomposition of a high N containing crop such as a legume, binding of the soil by the microbial cells themselves might be more important than binding by excretory products because N would not normally be limiting (Elliott & Lynch, 1984a).

Crop residue decomposition is controlled by water, temperature, chemical composition, and the associated macro- and microflora. Measurement and prediction of crop residue decomposition is critical to determine how long the residues will remain to protect the soil from erosion, to describe nutrient cycling, and to measure and manage the contribution of residues to soil structure and soil organic matter. A successful predictive effort must be mechanistic so that the model has wide application and the importance of the individual processes driving the process can be determined (Van Veen *et al.*, 1984). This approach provides accurate prediction of residue decomposition and the effects on soil structure, soil organic matter, and nutrient cycling. The tool also allows close examination of processes affecting these soil properties and management possibilities for improving beneficial effects of residues on soil properties. For example, studies by Knapp *et al.* (1983a) indicated that initial readily available carbon and nitrogen pools in wheat straw determined microbial biomass size and rate of straw decomposition. If nitrogen was limiting during utilization of the readily available carbon pool, extracellular polysaccharides were likely to be produced instead of microbial biomass. Reinertsen *et al.* (1984) postulated that the readily available C and N in wheat straw was not all soluble and that there appeared to be two readily available pools (one cold water soluble and one insoluble) that probably determined the size of the microbial biomass and resultant rate of decomposition. They also presented evidence that the N content of the two pools differed. These results confirm the weakness of using C:N ratios to predict residue decomposition rates. Parr & Papendick (1978) stated: "Although the N content or C:N ratio of crop residue can be useful in predicting decomposition rates, they should be used with some caution since the C:N ratio says nothing about the availability of the carbon or nitrogen to microorganisms."

Lynch & Elliott (1983) showed microbial biomass could stabilize soil and that when degraded wheat straw was added to the soil both the microbial biomass and gums appeared to be aggregating the soil. From the work of Knapp *et al.* (1983a, b) and Reinertsen *et al.* (1984), it can be postulated that extracellular materials such as polysaccharides might dominate the aggregation shown by decomposing straw if the wheat straw contains low N and if external N is unavailable. Elliott & Lynch (1984a) aerobically degraded three wheat straws containing 1.09, 0.5 and 0.25% N in the absence of added N. After a degradation period, the 0.25% N straw treatment aggregated the soils tested significantly greater than the other treatments, while the 0.5% N straw treatment generally caused more aggregation than the 1.09% N straw. The greatest microbial biomass would be generated from the highest N containing straw, and thus these results appear to confirm the postulate that the increased aggregation resulted from the production of extracellular gums. Presumably, the majority of these compounds are polysaccharides. Polysaccharides have long been implicated in soil aggregation (Tisdall & Oades, 1982). Lynch (1981) found the aggregate stabilizing ability of pure cultures of bacteria varied greatly. In many cases their ability to cause aggregation appeared related to their ability to produce extracellular polysaccharides on solid media; but this was not always so. The results of these studies show that there is a potential for improving soil aggregation through residue management on the soil surface. Improved soil aggregation increases water infiltration, resistance to erosion, and presumably soil productivity.

Soil organic matter

The importance of soil organic matter to soil physical, chemical, and biological properties with implications on erodibility, nutrient cycling, water storage, plant vigor, and resultant soil productivity is well established (Larson *et al.*, 1972; Martin & Stott, 1984; Stevenson, 1982). Results have shown that management and rate of residue addition to soils can affect the amount of soil organic matter and its distribution. Several studies indicate that direct seeding of crops reduces loss of soil organic matter through erosion and the decomposition stimulated by tillage (Lal, 1974; Dick, 1984; Bauer & Black, 1981).

We feel there is opportunity to manipulate crop residue decomposition so that the more active soil organic matter fractions are increased. First it will be helpful to review the relationships of soil organic matter to crop residue and tillage management. Rovira & Greacen (1957) pointed out that tillage increases the rate of soil organic matter mineralization. Hiura *et al.* (1976) found a strong correlation between the increased mineralization of N with grinding and the clay/humus ratios. In his review, Stevenson (1982) pointed out that cultivation reduces soil organic matter and N. Soil organic matter levels can be maintained by inclusion of a sod crop in the rotation or by the frequent addition of large quantities of organic residues. Martin & Stott (1984) agreed with these statements. They also pointed out that humus decomposes at the rate of 2 to 5% per year and that new humus decomposes much more rapidly than old with 99% of the new humus turning over in a time frame of 25–60 years. This active humus or soil organic matter fraction is the one we are most concerned with because it should be most actively involved in the beneficial changes to soil properties. When soil organic matter content is increased, presumably the bulk of change would be associated with this active fraction. As this fraction size increases, the size of the microbial biomass increases with beneficial changes on soil structure and nutrient retention and release.

While the picture is not clear, the surface management of crop residues and direct seeding of the following crop or establishment of the following crop with minimal soil tillage appears to have the greatest promise for improving the quality of soil organic matter.

In the northern Great Plains of the U.S., Bauer & Black (1981) found the average organic C and total N to a depth of 45.7 cm (percent by weight) were significantly higher under stubble mulch than conventional tillage in a fine- and coarse-textured soil. They attributed the difference to better erosion control by the stubble mulch tillage. Studies on native prairie cropped by no-till and conventional till seeding in western Nebraska showed that soil N and organic C remained about the same as the native prairie when no-till seeded while it significantly decreased for the conventional tillage treatments (Lamb *et al.*, 1985). They suggested decreased soil stirring resulted in N conservation. Much earlier, Lal (1974) obtained similar results in western Nigeria.

Reports on comparisons of the effect of different tillage on soil organic matter on sites that had previous histories of long-term tillage present a mixed picture. Dick (1984) installed tillage trials following a corn-oats-meadow rotation and found at the Hoytsville, Ohio, site that after 18 years the organic C had not changed in the no-till plots while organic C decreased 12 and 14% in the companion minimum till and conventional till cropped plots, respectively. At the Wooster, Ohio, site he installed tillage trials in a grass meadow. After 18 years of cropping, organic C was reduced 11, 23, and 25% in the soils no-till, minimum till, and conventional till cropped, respectively, as compared with organic C remaining in the grass meadow soil. Soil organic N content followed similar patterns.

Doran (1980) surveyed several tillage trials across the U.S. and found microbial numbers and soil organic matter were greater near the surface of no-till trials but, when averaged over the plow depth, changes were not apparent. Powlson & Jenkinson (1981) concluded that changing from traditional cultivation methods had little effect on soil organic matter other than altering its distribution in the profile. Carter & Rennie (1982) studied soil biological changes affected by tillage and found no significant differences in total soil organic C and N between no-till and conventional cropping after 16 years; however, they did find differences in distribution. The same trends were evident for soil microbial biomass. Lynch & Panting (1982) obtained similar results. While total differences were difficult to show in these studies, the differences in distribution of soil biological properties affected by tillage could be most significant, especially near the soil surface. The soil surface controls erodibility, infiltration, and evaporation. This area could also influence nutrient flow and early plant vigor.

It is well known that the addition of residues to soil increases humus (Martin & Stott, 1984; Stevenson, 1982). The inclusion of a sod crop in the rotation also increases soil organic matter (Stevenson, 1982). Legumes, green manures, or barnyard manure in the cropping sequence slows the loss of C and N from the soil (Haas *et al.*, 1957). Larson *et al.* (1972) demonstrated the importance of returning crop residues to soil and the benefits on soil C. When crop residues were removed by repeated burning, Biederbeck *et al.* (1980) found bacterial populations were permanently reduced by more than 50% in the top 2.5 cm of soil while the fungi appeared to recover. The burning also reduced soil N and C and potentially mineralizable N in the 0–15 cm segment. Burning residues appears to reduce soil organic matter, reduce water stable aggregates, and increase peak discharge rates of runoff (Dormaar *et al.*, 1979; Marston, 1978; Unger *et al.*, 1973). Organic farming methods are credited with less erosion presumably because of an active soil organic matter fraction (Papendick & Elliott, 1984). Lal *et al.* (1980) demonstrated the benefits of residues on the surface by comparing plots mulched and nonmulched with rice straw. They found that the mulch greatly impeded the detrimental changes in soil physical condition and chemical relationships in the 0–5 cm layer.

While the picture is not clear, surface-managed residues beneficially affect the upper portion of the profile. Possibly after long term studies it will emerge that the entire profile benefits. The soil surface is a very important portion of the profile. The beneficial effects of using the crop residues and green manures and pastures in the rotation is well documented. However, the best methods for management are not.

Plant growth and vigor

The effect of residues and decomposing residues on plant growth and vigor has been widely studied. Crop growth problems have been associated with a variety of residues worldwide mainly when crops are direct-seeded into the residues or seeded into shallow-incorporated residues (Cannell *et al.*, 1982; Elliott *et al.*, 1980; Jessop & Stewart, 1983; Lovett *et al.*, 1982; Lynch *et al.*, 1981). The problems occur primarily when large quantities of crop residues are produced and the following crop is direct seeded into the relatively fresh residues or the residues are shallow incorporated into the soil. The problem is severe enough that there have been attempts to incorporate these principles for weed control (Steinsiek *et al.*, 1982). These problems have resulted in efforts to dispose of the residues by burning or other methods which result in good crop growth. However, nutrient costs and replaceability, soil erodibility, and maintaining good soil structure demand that we utilize the crop residue resource.

Plant diseases and residue management will not be discussed here as they are covered elsewhere (see R.J. cook, this volume).

Some of the severe plant growth problems that occur when crops are seeded into surface-managed residues in areas with winter rainfall climates such as England and the Pacific Northwest are attributed to the production of the short-chain fatty acids, acetic and butyric (Lynch *et al.*, 1980; Lynch & Elliott, 1984; Elliott *et al.*, 1984). The acids do not travel far through the soil so the roots of the growing plant must almost touch the residues to be affected. The short-chain fatty acids appear to cause problems only during cool-wet falls. Management of residues away from the seed row, better residue spreading, and good soil-seed contact should solve this problem. Complete residue removal also solves the problem (Cannell *et al.*, 1982), but the nutrient and conditioning value of the residues is lost.

The chaff (flowering remains of the seed head) remaining after harvest appears to be a much greater problem than the straw. It appears to be a serious problem every year (personal observations). The chaff is difficult to spread, especially behind large combines, and rarely do vigorous healthy grain stands result where direct seeded into the areas of concentrated chaff. In fact, weeds fail to grow also.

Another area that may be associated with the residue management problem is that winter wheat (*Triticum aestivum* L.) roots can be heavily colonized by pseudomonad bacteria that are very inhibitory to wheat root growth in laboratory tests. Wheat cultivars responded differently to the effects of these organisms (Elliott & Lynch, 1984b). The organisms appear related to heavy cereal residues but we have been unable to establish direct correlations. They appear on the roots in large numbers after the wheat breaks dormancy in late winter with as high as 100% of the isolates tested in the laboratory being inhibitory to seedling wheat root growth. The organisms appear to produce a toxin that adversely affects the plant growth and they colonize the roots aggressively even when they have mutated so they do not impede plant root growth. Few root abnormalities are caused by these organisms. Retardation of root growth and some retardation of shoot growth are the main visible effects (Elliott & Lynch, 1984b, 1985; Fredrickson & Elliott, 1985a,b). The effect of these organisms on wheat yield is conjectural at this point. The information indicates they could be an important plant health factor. Simple residue management may not solve this problem. Improved varieties, development of inocula, improved drilling practices, or better and more frequent crop rotations may be the most viable approaches. These organisms may also have the ability to affect early plant vigor. As stated earlier, plant vigor is the best method of sustaining crop production.

While we can point to many negative aspects on plant growth associated with surface-managed residues, positive information is emerging also. Doran *et al.* (1984) studied dryland production of no-till corn (*Zea mays* L.), sorghum (*Sorghum bicolor* L. Moench), and soybean (*Glycine max* L.) and the effect of crop residue levels on their production for 3 years. They found that when the previous crop residues were completely removed after harvest, grain and residue yields of corn and soybean were 22 and 24% lower, respectively, compared with no residual removal. They attributed the yield reductions to temperature and moisture stress. Wilhelm *et al.* (1985) continued the experiment and found there was a significant linear response between grain and residue yield and amount of previous crop residue applied to the soil surface (0, 50, 100, and 150%). Again, they attributed the positive effect of the residues to water conservation and moderation of soil temperature.

Drilling practices and plant geometry

The greatest problem we generally face with direct seeding through crop residues is drilling equipment that will place the seed in good contact with the soil, cut through heavy residues without tucking, manage the straw away from the seed row, and properly place nutrients (if needed).

While direct drilling into surface residues may seem incompatible with organic farming methods, the advantages of erosion control, positive effects on soil surface structure, the possibilities for increasing microbial biomass and efficiency of nutrient cycling, and decreasing the deleterious effects of tillage on soil organic matter require that we look at this approach for a sustainable farming system.

The recent development of direct seeding equipment by USDA-ARS personnel (Pullman, Wash.) in cooperation with private industry, has provided direct drilling capabilities that we did not have in the past. The drill penetrates residues, assures good soil-seed contact, and, if used, can place various fertilizers with the seed and/or between the seed row. The machine uses what is called the "paired row" concept. Two seed rows are placed 12.7 cm apart to form a "pair" and, if desired, fertilizer can be banded between the "pair" to a depth of 11.4–12.7 cm. The pairs of rows are spaced 38.1 cm apart (Fig. 1). The geometry of the pairs of rows appears to stimulate plant growth. Field observations on fertilized and nonfertilized trials (no fertilizers of any kind for four years except two green manure crops) show that the paired row seeding was more vigorous throughout the growing season than companion conventionally tilled and seeded trials (Elliott & Papendick, unpublished results). This trend is surprising because direct seeded plots stay colder in spring than tilled areas because reflectance is greater where residues are present.

The apparent increase in plant vigor in areas direct seeded with the "paired row" machine may result from alleviation of toxic problems from the residues, less root colonization by inhibitory microbes, fewer pathogen problems, or other factors of which we are not aware.

The "paired row" direct seeding concept has interesting implications. The wide space between the "pairs" allows row crop weed control either by shallow cultivation or shielded spraying. In our observations, weeds are seldom a problem between the rows spaced 12.7 cm apart because the vigorous crop growth crowds the weeds. The row spacing may not be optimum, but it appears to work well in the Pacific Northwest. In the northern Great Plains of the U.S., spacing between the two rows of 17.8 cm with 33 cm between the pairs seems to work well (personal observations). The spacing between the "pair" was widened to facilitate seeding wet soil. However, this spacing would not lend itself as well to row crop weed control.

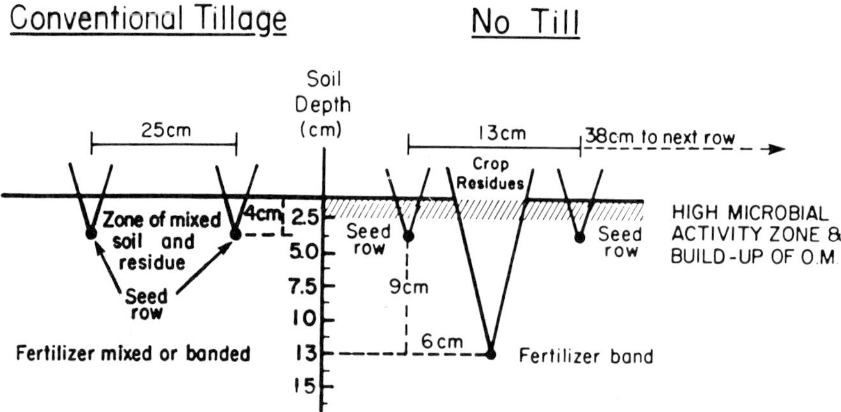

FIGURE 1   Paired row seeding configuration compared with conventional.

While synthetic or manufactured fertilizers are not usually used in standard organic farming practices, the "paired row" direct seeding concept may lend itself well to a minimal use of these fertilizer sources. Small quantities of fertilizer with the seed as a "starter", especially in cold wet ground, could be very beneficial. A small quantity of N banded between the rows constituting a pair could be used theoretically at good efficiency and the weeds would not benefit from this application.

Biological activity in organic and conventionally farmed soil

Many statements have been made that organic farming techniques are beneficial to the soil microflora, yet few data exist. At Pullman selected microbiological properties of soils were studied from adjoining wheat farms in the highly productive Palouse region of eastern Washington. Since 1909, the only N input to the soil of the organic farm has been through leguminous green manure crops consisting most recently of Austrian winter peas (*Pisum sativum* ssp. *arvense* L. Poir) about once every 4–5 years, plus native soil fertility for N and all other plant nutrients. The organic farm is 320 hectares. Comparatively, the soil of the inorganic farm received regular applications of anhydrous ammonia, P, and S at recommended rates for the last 30 years. Not surprising, there were no differences in numbers of soil microorganisms as determined by plate counts; however, the soil from the organic farm had significantly higher levels of urease, phosphatase, and dehydrogenase at all three samplings and significantly higher soil microbial biomass at the first two samplings. The data indicate that the organic farm soil had a larger and more active soil microflora (Bolton *et al.*, 1985). In another study it was shown that the organic farm still had very adequate available phosphorus; in fact, available P tended to be higher in the 10–15cm soil depth (A.W. Patten and R.I. Papendick, unpublished results). Crop yields for the organic farm compared favorably with the neighbor's yield obtained with synthetic fertilizers; however, the organic farm was not as intensively cropped because of the green manure crop in the rotation.

These results and the USDA survey conclusively show that organic farms of appreciable size can be operated successfully (USDA, 1980) and that it does appear to stimulate the soil microflora. It is questionable that the limited use of green manure was solely responsible for the greater microbial activity in the soil from the organic farm in the Palouse region.

SUMMARY

Residue management appears to be very important for maintaining and improving soil structure and soil organic matter. Surface management of residues with reduced tillage or direct seeding appears to be the most viable approach to achieve these goals and possibly increase soil productivity rather than sustaining it or slowing the loss of it. Drilling equipment has been a problem, but the development of the "paired row" seeding concept appears to be a major breakthrough. Mating of this concept with organic farming appears to be a viable approach with no or very limited use of synthetic fertilizers. Row crop weed control is viable with this system. Also, organic farming practices appear to stimulate the soil microflora.

With no-till seeding, toxicity problems can result from products produced during residue decay. It appears these problems can be alleviated by managing residues out of the seedling row. Root colonization by inhibitory pseudomonads could be a serious problem, but it is too early to determine. We find them on the plant root surface in large numbers, and the organisms cause severe plant growth retardation in the laboratory and greenhouse. If they are a severe problem in the field, residue management, rotations, and plant variety development should alleviate the problem.

References

Bauer, A. & Black, A.L. (1981). Soil carbon, nitrogen, and bulk density comparisons in two cropland tillage systems after 25 years and in virgin grassland. *Soil Science Society of America Journal*, **45**, 1160–1170.

Biederbeck, V.O., Campbell, C.A., Bowren, K.E., Schnitzer, M. & Medner, R.N. (1980). Effect of burning cereal straw on soil properties and grain yields inSaskatchewan. *Soil Science Society of America Journal*, **44**, 103–111.

Bolton, H., Jr., Elliott, L.F., Papendick, R.I. & Bezdicek, D.F. (1985). Soil microbial biomass and selected soil enzyme activities: Effect of fertilization and cropping practices. *Soil Biology & Biochemistry*, **17**, 297–302.

Cannell, R.Q., Ellis, F.B., Christian, D.G. & Barnes, B.T. (1982). Long-term comparisons of direct drilling, shallow tillage and ploughing on clay and silt-loam soils, with particular reference to straw disposal. In *The 9th Conference of the International Soil Tillage Research Organization*, ISTRO, Socialistic Republic of Yugoslavia, Osijek. Pp.85–90.

Carter, M.R. & Rennie, D.A. (1982). Changes in soil quality under zero tillage farming systems: Distribution of microbial biomass and mineralizable C and N potentials. *Canadian Journal of Soil Science*, **62**, 587–597.

Dick, W.A. (1984). Organic carbon, nitrogen, and phosphorus concentrations and pH in soil profiles as affected by tillage intensity. *Soil Science Society of America Journal*, **47**, 102–117.

Doran, J.W. (1980). Soil microbial and biochemical changes associated with reduced tillage. *Soil Science Society of America Journal*, **44**, 765–771.

Doran, J.W., Wilhelm, W.W. & Power, J.F. (1984). Crop residue removal and soil productivity with no-till corn, sorghum, and soybean. *Soil Science Society of American Journal*, **48**, 640–645.

Dormaar, J.F., Pittman, U.J. & Spratt, E.D. (1979). Burning crop residues: Effect on selected soil characteristics and long-term wheat yields. *Canadian Journal of Soil Science*, **59**, 79–86.

Elliott, L.F., Cochran, V.L. & Papendick, R.I. (1980). The effect of crop residue management on phytotoxicity, straw decomposition, and winter wheat root colonization. In *Proceedings of the Tillage Symposium*, Bismark, North Dakota, September 9–11, pp.88–96. Cooperative Extension Service, North Dakota State University; Fargo.

Elliott, L.F. & Lynch, J.M. (1984a). The effect of available carbon and nitrogen in straw on soil and ash aggregation and acetic acid production. *Plant & Soil*, **78**, 335–343.

Elliott, L.F. & Lynch, J.M. (1984b). Pseudomonads as a factor in the growth of winter wheat (*Triticum aestivum* L.). *Soil Biology & Biochemistry*, **16**, 69–71.

Elliott, L.F. & Lynch, J.M. (1985). Plant growth-inhibitory pseudomonads colonizing winter wheat (*Triticum aestivum* L.) roots. *Plant & Soil*, **84**, 57–65.

Elliott, L.F., Papendick, R.I. & Cochran, V.L. (1984). Phytotoxicity and microbial effects on cereal growth. In sixth Manitoba-North Dakota Zero Tillage Workshop, Bismark, North Dakota, January 26 and 27, pp.40–47. Cooperative Extensive Service, North Dakota State University; Fargo.

Fredrickson, J.K. & Elliott, L.F. (1985a). Effects on winter wheat seedling growth by toxin-producing rhizobacteria. *Plant & Soil*, **83**, 399–409.

Fredrickson, J.K. & Elliott, L.F. (1985b). Colonization of winter wheat roots by inhibitory rhizobacteria. *Soil Science Society of America Journal*, **49**, 1172–1177.

Gilmour, C.M., Allen, O.N. & Truog, E. (1948). Soil aggregation as influenced by the growth of mold species, kind of soil, and organic matter. *Soil Science Society of America Proceedings*, **13**, 291–296.

Haas, H.J., Evans, C.E. & Miles, E.F. (1957). Nitrogen and carbon changes in Great Plains soils as influenced by cropping and soil treatment. Technical Bulletin 1164, USDA. U.S. Government Printing Office; Washington, D.C.

Hayes, M.H.B. & Swift, R.S. (1978). The chemistry of soil organic colloids. In *The Chemistry of Soil Constitutents* (D.J. Greenland & M.H.B. Hayes, eds), pp.179–320. John Wiley & Sons; New York.

Hiura, K., Hattori, T. & Furusaka, C. (1976). Bacteriological studies on the mineralization of organic nitrogen in paddy soils. II. Effect of mechanical disruption of soils on ammonification and bacterial number. *Soil Science & Plant Nutrition*, **22**, 459–465.

Jessop, R.S. & Stewart, L.W. (1983). Effects of crop residues, soil type and temperature on emergence and early growth of wheat. *Plant & Soil*, **74**, 101–109.

Knapp, E.B., Elliott, L.F. & Campbell, G.S. (1983a). Microbial respiration and growth during the decomposition of wheat straw. *Soil Biology & Biochemistry*, **15**, 319–323.

Knapp, E.B., Elliott, L.F. & Campbell, G.S. (1983b). The interrelationships of carbon, nitrogen and microbial biomass during the decomposition of wheat straw: A mechanistic simulation model. *Soil Biology & Biochemistry*, **15**, 455–461.

Lal, R. (1974). No-tillage effects on soil properties and maize (*Zea mays* L.) production in western Nigeria. *Plant & Soil*, **40**, 321–331.

Lal, R., De Vleeschauwer, D. & Nganji, Malafa R. (1980). Changes in properties of a newly cleared tropical Alfisol as affected by mulching. *Soil Science Society of America Proceedings*, **44**, 827–833.

Lamb, John A., Peterson, G.A. & Fenster, C.R. (1985). Wheat fallow tillage systems' effect on a newly cultivated grassland soils nitrogen budget. *Soil Science Society of America Journal*, **49**, 352–356.

Larson, W.E., Clapp, C.E., Pierre, W.H. & Morachan, Y.B. (1972). Effect of increasing amounts of organic residues on continuous corn: organic carbon, nitrogen, phosphorus, and sulfur. *Agronomy Journal*, **64**, 204–208.

Low, A.J. (1972). The effects of cultivation on the structure and other characteristics of grassland and arable soils (1945–1970). *Journal of Soil Science*, **32**, 363–380.

Lovett, J.V., Hault, E.H., Jessop, R.S. & Purvis, C.E. (1982). Implications of stubble retention. *Proceedings of the Second Australian Agronomy Conference*, pp.101–115.

Lynch, J.M. (1981). Promotion and inhibition of soil aggregate stabilization by micro-organisms. *Journal of General Microbiology*, **126**, 371–375.

Lynch, J.M. (1984). Interactions between biological processes, cultivation and soil structure. *Plant & Soil*, **76**, 307–318.

Lynch, J.M. & Elliott, L.F. (1983). Aggregate stabilization of volcanic ash and soil during microbial degradation of straw. *Applied & Environmental Microbiology*, **45**, 1398–1401.

Lynch, J.M. & Elliott, L.F. (1984). Crop Residues. In *Crop Establishment: Biological Requirements and Engineering Solutions* (M.K.V. Carr, ed.). Pitman; London. (In press).

Lynch, J.M., Ellis, F.B., Harper, S.H.T. & Christian, D.G. (1981). The effect of straw on the establishment and growth of winter cereals. *Agriculture & Environment*, **5**, 321–328.

Lynch, J.M. & Panting, L.M. (1982). Effect of season, cultivation, and nitrogen fertilizer on the size of the soil microbial biomass. *Journal of the Science of Food & Agriculture,* **33**, 249–252.

Marston, D. (1978). The effect of two stubble management practices on runoff characteristics in arable black earth soils. In *Proceedings of the Technical Conference of the Institute of Australian Engineers,* Toowoomba, Queensland, Australia, 29–31 August, 1978. Pp.130–133.

Martin, J.P. (1971). Decomposition and binding action of polysaccharides in soil. *Soil Biology & Biochemistry,* **3**, 33–41.

Martin, J.P.& Stott, D.E. (1984). Characteristics of organic matter decomposition and retention in arid soils. In *Sixth Annual International Symposium for Environmental Biogeochemistry* (J.J. Skujins, ed). UNEP Publisher; Paris.

Parr, J.F. & Papendick, R.I. (1978). Factors affecting the decomposition of crop residues by microorganisms. In *Crop Residue Management Systems* (W.R. Oschwald, ed.), pp. 109–129. American Society of Agronomy; Madison, WI.

Papendick, R.I. & Elliott, L.F. (1984). Tillage and cropping systems for erosion control and efficient nutrient utilization. In *Organic Farming: Current Technology and Its Role in Sustainable Agriculture* (D.F. Bezdicek, J.F. Power, D.R. Keeney & M.J. Wright, eds), pp.69–81. American Society of Agronomy; Madison, WI.

Powlson, D.S. & Jenkinson, D.S. (1981). A comparison of the organic matter, biomass, adenosine triphosphate and mineralizable nitrogen contents of ploughed and direct-drilled soils. *Journal of Agricultural Science, Cambridge,* **97**, 713–721.

Reinertsen, S.A., Elliott, L.F., Cochran, V.L. & Campbell, G.S. (1984). Role of available carbon and nitrogen in determining the rate of wheat straw decomposition. *Soil Biology & Biochemistry,* **16**, 459–464.

Rovira, A.D. & Greacen, E.L. (1957). The effect of aggregate disruption on the activity of microorganisms in the soil. *Australian Journal of Agricultural Research,* **8**, 659–673.

Steinsiek, J.W., Lawrence, O.R. & Collins, F.C. (1982). Allelopathic potential of wheat (*Triticum aestivum*) straw on selected weed species. *Weed Science,* **30**, 495–497.

Stevenson, F.J. (1982). *Humus Chemistry, Genesis, Composition, Reactions.* John Wiley & Sons; New York. 443pp.

Tisdall, J.M. & Oades, J.M. (1982). Organic matter and water-stable aggregates in soils. *Journal of Soil Science,* **33**, 141–163.

Unger, P.W., Allen, R.R. & Parker, J.J. (1973). Cultural practices for irrigated winter wheat production. *Soil Science Society of America Proceedings,* **37**, 437–442.

U.S. Department of Agriculture (USDA). (1980). Report and recommendations on organic farming. A special report prepared for the Secretary of Agriculture. U.S. Government Printing Office; Washington, D.C. 164pp.

Van Veen, J.R., Ladd, J.N. & Frissel, M. (1984). Modelling C and N turnover through the microbial biomass in soil. *Plant & Soil,* **76**, 257–274

Wilhelm, W.W., Doran, J.W. & Power, J.F. (1986). Corn and soybean yield response to crop residue management with no tillage. *Agronomy Journal.* In press.

Rhizosphere Microbiology and its Manipulation

J.M. Lynch

Glasshouse Crops Research Institute, Littlehampton, West Sussex, BN17 6LP, U.K.

INTRODUCTION

The first definition of the rhizosphere by Hiltner in 1904 concerned the stimulation of microbial growth around legume roots but it was shortly after this that the concept was applied to all crops. For a similar length of time it has been recognised that dead roots and other plant residues provide the other major source of carbon and energy substrates for the soil microflora. The link between the two microbial populations has perhaps been less obvious; inevitably if a plant root grows through soil where a plant residue is being decomposed its rhizosphere population is likely to be modified by the community degrading the residue and certainly any metabolites formed will potentially have an influence on the root. In this respect therefore it is probably only reasonable to view both the microbial population growing directly on root-derived carbon and the population degrading residues in the vicinity of plant roots as part of a rhizosphere continuum. The continuum should also include the microbial populations entering the root cortex (Darbyshire & Greaves, 1973; Old & Nicolson, 1978) and the interaction between the rhizosphere of roots of the same and different crop species (Newman, 1985).

The microbial population of the rhizosphere includes saprophytes and mutualistic and antagonistic symbionts (Whipps & Lynch, 1985). Thus whereas it is perhaps the most complex region of soil in terms of the population and community dynamics, it is also the region of soil which will have the most profound influence on crop productivity and is therefore of most relevance to a discussion of sustainable agriculture. Indeed, crops produced commercially in hydroponic and other forms of soil-less culture come totally under the influence of a rhizosphere population. The central point for discussion is how much of the plant's nutrient and other needs can be provided by the rhizosphere population while avoiding the harmful effects of micro-organisms. What is the scope for manipulation of the population by elevating the proportion of beneficial against the harmful members? Before that is considered, the current state of knowledge on provision and utilisation of energy sources in the rhizosphere will be considered.

ENERGY SOURCES AND POPULATION DYNAMICS OF THE RHIZOSPHERE

The rhizosphere is a continuum which starts immediately around the root (the ectorhizosphere), includes the root surface (the rhizoplane) and continues into the root cortex (the endorhizosphere). In the early studies of carbon release by roots to micro-organisms the soluble fraction which is actively pumped (secretions), leaked (exudates) or passively released (lysates) from roots was the primary consideration. However, that material may only be a small fraction of that available to micro-organisms, the remainder being the mucigel (plant and microbial mucilage) which is largely polysaccharide in nature. Those materials can only be collected from axenic plants growing in soil-less culture, which may well produce a different pattern of carbon release compared with plants growing naturally, and even then the total fraction is difficult to collect. To overcome this difficulty carbon release has been studied by growing plants in an atmosphere of $^{14}CO_2$ with the root system in a sealed container of soil and measuring the ^{14}C entering the soil as solubles, insolubles and respired $^{14}CO_2$ (Barber & Martin, 1976). Early studies showed that this carbon release could account for up to 20% of the plant's photosynthate produced but more recent studies (Whipps & Lynch, 1983; Whipps, 1984) have shown this can be around 30%. This has immediate implications for sustainable agriculture in that there is a large apparent wastage in plant biomass production which is channelled directly to the rhizosphere microflora. Is this a functionless waste or does the plant get a dividend from the microflora?

In the microbial utilization of the substrate provided by roots, a critical factor is the C:N ratio. Microbial cells usually have a C:N ratio of about 5:1. There is hardly any information on the C:N ratios of the carbon released by roots but as there is a substantial amount of polysaccharide it seems likely to be of the order of 40:1. If it is assumed that all the carbon and nitrogen is available to the micro-organisms and that 35% of the C used (the remainder going to CO_2) and 100% of N used enters the biomass, it is clear that inadequate N is available to satisfy a microbial biomass with a C:N ratio of 5:1. If 100 g of root-derived carbon contained 40 g C and 1 g N, then an extra 1.8 g of N would need to be provided from soil or fertilizer N to satisfy the decomposition requirement. However, this calculation is a gross oversimplication because the C:N ratios have not been determined. Furthermore not all of the carbon is available to micro-organisms, particularly the mucigel, and even then the N immobilized in micro-organisms will be recycled by predatory protozoa and other micro-organisms. The recycled N may be taken up directly by the plant root and indeed Clarholm (1983) has provided evidence that this may be an important route for NH_4^+ uptake by the plant. Clearly this is an area of great relevance to reduced-input farming systems and where more research effort should be directed.

Many species of bacteria, fungi and protozoa co-exist in the rhizosphere but a fundamental question is to understand which species take the largest share of the available substrate because those are most likely to have an influence on plant growth. The colonization potential is the number or biomass of the particular species which can be supported per unit length or weight of root (Bennett & Lynch, 1981a). This can be determined by each particular species in axenic gnotobiotic culture individually and in mixed culture. The concept can be applied to the soil environment and is particularly relevant to the introduction of biocontrol agents to control root pathogens. The effect of such colonization on plant growth must also be considered both in the absence and in the presence of plant pathogens. A problem in interpreting such results is that whereas the colonization potential may be constant between experiments, the effect on plant growth can be variable; this will be discussed further later. Gnotobiotic studies where a second or more organism(s) are simultaneously introduced to the plant root are useful in assessing the competitive ability linked to the colonization potential (Bennett & Lynch, 1981b). Inevitably the growth of the plant in nonsterile soil will be more complex, but a 'screen' or 'bioassay' of this type will help to produce a candidate list of useful micro-organisms to introduce to the rhizosphere of a particular crop. This will be especially relevant where the competing organism in the screen is a pathogen. From such empirical studies, glasshouse trials in pots of soil can follow. Here it may be useful to produce antibiotic-resistant strains of the introduced organism so that it can be recovered and counted on agar containing antibiotic. However, ultimately the efficacy of the introduced organism must be tested in field soil. Nevertheless, it seems that it will ultimately be more economical in time and labour to screen potential inoculants fully in the laboratory and glasshouse such that their effects are consistent in several successive experiments before becoming involved in more labour-intensive field experiments. This procedure is not foolproof in that some isolates useful in the field will fail such tests and *vice versa* but this is a problem that industry has come to live with over the years in the screening of potential agrochemicals.

CONSEQUENCES OF THE RHIZOSPHERE FOR THE PLANT

The colonization of plant roots by symbionts or saprophytes can lead to beneficial or harmful effects on the plant. The positive effects can include direct effects, such as the production of plant growth stimulators and the promotion in plant nutrient uptake, and indirect effects. The latter include the production of antibiotics against pathogenic species, modifying the biological population present in soil and the production of soil conditioners, modifying the soil physical environment to make it more favourable for plant growth.

Production of plant growth regulators

In the early attempts at seed inoculation, largely in the Soviet Union, it was suggested that the preparations 'azotobacterin' and 'phosphobacterin' acted by providing the plant root with N and P (Mishustin, 1970). Similarly in the more recent attempts at inoculation with N_2-fixing bacteria, particularly *Azospirillum* spp., isolated from the roots of tropical grasses, it was considered initially that the inoculum functioned by providing N to the plant (Dobereiner & Day, 1976). The nitrogenase enzyme system is very inefficient at N_2-fixation in energetic terms and a theoretical analysis shows that it is unlikely that N_2-fixation in the rhizosphere is likely to be significant to crop plants but could be significant in their ecological survival (Barber & Lynch, 1977; Gilmour & Gilmour, 1978). From the many investigations worldwide it is clear that there have been many genuine reports where microbial inoculants of seeds have stimulated plant growth. However, responses have generally been inconsistent. Early demonstrations that *Azotobacter chroococcum* can produce metabolites which show positive bioassay results towards gibberellins have led Brown (1974) to suggest that the production of plant growth regulators is the likely mode of action of seed inoculants. However, the chemical characterization of the substances giving the positive bioassays has generally not been attempted. Rhizosphere microbiology will not advance greatly until this type of challenge has been met. With the new opportunities created by biotechnology, it would be possible to genetically manipulate the production of such substances but this depends on definite characterization. Even then, however, it is critical to be certain of the effect of the metabolite on the physiology of the plant. The role of endogenous plant growth regulators is relatively unequivocal and therefore if micro-organisms produce these substances in the endorhizosphere, their effect on the plant can be predicted. However, it is likely that the bulk of such metabolites would be produced in the ectorhizosphere or on the rhizoplane. Here their effect is less predictable and indeed in a study using axenic cereal plants (Lynch & White, 1977) the roots appeared rather insensitive to exogenous regulators.

If plant growth regulators do not affect roots directly, it is possible that they could have an indirect effect by affecting nutrient uptake processes. This might only happen in nutrient-poor environments and would therefore be difficult to demonstrate in luxuriant nutrient media in the laboratory. However, these ideas are total speculation and there is a clear need for detailed investigations. Such studies are essential to bring scientific credibility to rhizosphere investigations.

Plant Nutrient Uptake

By contrast to the studies on seed inoculation, work on the role of micro-organisms in nutrient uptake by plants has been mainly in well-defined nutrient culture systems in the laboratory and little has been attempted in soil. Of the symbionts, mycorrhiza have been studied extensively in demonstrating their role in increasing phosphorus uptake by the plant and this will be reviewed by Mosse elsewhere in this volume. However, it is perhaps salutory to consider if 'something for nothing' is gained in these associations. The symbiont is a carbon drain on the plant and an increase in the P/C ratio of the plant as a result of mycorrhizal colonization is not necessarily a net benefit. However, carbon does not appear to be the major limitation on plant growth.

The studies with saprophyte effects on nutrients were done mainly in the CSIRO Adelaide and AFRC Letcombe Laboratories in the period 1966–76. Contrasting reports over the years were resolved by demonstrating that over short (30 min) periods phosphate uptake and translocation within the barley plant are promoted by bacteria but over longer (24 h) periods uptake is reduced (Barber *et al.,* 1970). Seedling age is also important, phosphate uptake being stimulated by the bacteria in younger (6d) seedlings and inhibited in older (12d) seedlings. Other studies included a demonstration on the promotion of manganese uptake by bacterial chelating agents (Barber & Lee, 1974). There is clearly a need for soil experiments in this area because the effects are highly relevant to low-input farming systems.

Kloepper *et al.* (1980) have demonstrated that some bacteria produce iron-chelating agents (siderophores) and they can restrict the availability of iron to plant pathogens. This is only useful if the chelated iron becomes available to the plant, for which there is limited evidence (Powell *et al.,* 1982). It must also be emphasized that the capacity of soil bacteria and fungi to produce siderophores is widespread (Focht & Verstraete, 1977; Powell *et al.,* 1980; Szaniszlo *et al.,* 1981) and therefore inoculation for this function may be of limited benefit.

Antibiosis

Since the early studies of Selman Waksman on the production of antibiotic substances in soils by micro-organisms, the potential to harness this for the control of plant pathogens has been a tantalising prospect. Studies by the ICI group at the Frythe Laboratory demonstrated that a range of antibiotics could be formed in soils but Brian (1957) was in doubt about the ecological significance. He pointed out that, because the substances could be readily degraded or inactivated on soil colloids, they would probably only accumulate where there was an abundant supply of substrates. Decomposing plant residues was the primary site considered but the rhizosphere or the associated region around seeds, the spermosphere, is probably a better illustration of a region with abundant substrate supply. It is perhaps therefore somewhat surprising that the rhizosphere has not been investigated in detail as a source of antibiotic, and hence biocontrol, substances. However, it could be argued that in attempts to find biocontrol agents (see Cook, this volume), antibiotic production is an integral part of the action. For example Dennis & Webster (1971) demonstrated that a peptide material and a coconut-smelling volatile were produced by *Trichoderma* spp. and these appeared to be responsible for the biocontrol properties of the antagonist. Today, chemical characterization of such metabolites by the various spectroscopic methods does not present a major problem and indeed Dr N. Claydon in my laboratory has recently been able to characterize the *Trichoderma* volatile. There is scope to physiologically and genetically manipulate the production of such secondary metabolites and this should open a new exciting era of biocontrol. It should be recognized however that antibiotic production is unlikely to be the only mode of action of antagonists and indeed *Trichoderma* spp. produce a useful complement of cell-wall degrading enzymes (Elad *et al.,* 1982). There is also evidence of mycoparasitism by the antagonist coiling around the hyphae of the pathogen (Elad *et al.,* 1983).

The rhizosphere is a particularly useful target for biocontrol studies because there are few useful agrochemicals which can be applied to control root pathogens. Application to field crops is clearly the goal, but protected crops grown in relatively well-controlled environments offer greater prospect to obtain consistent effects, which is essential for credibility to be obtained for the approach. For example, the control of black root rot of cucumber, *Phomopsis sclerotioides,* have been admirably demonstrated by Ebben & Spencer (1978).

Soil Conditions

Micro-organisms produce a range of polysaccharides and other gums which can aggregate soil particles and prevent erosion (Lynch & Bragg, 1985). The magnitude of the effect depends on the size and composition of the microbial biomass present. Clearly the carbon released by roots provides a source for the generation of that biomass. The polysaccharides produced by roots themselves appear less satisfactory in aggregating than the rhizosphere biomass because sterile plants only have a small effect on the soil (E. Bragg, unpublished). Reid & Goss (1981) observed that whereas the ryegrass rhizosphere was effective in stabilizing structure, destabilizing effects were observed when maize was grown in the same soil. However, in other experiments (E. Bragg, unpublished) positive effects were found from both plant species and presumably there must have been some difference in experimental procedure. It should be recognized of course that whereas effects of polysaccharides are important and contribute perhaps 50% of the physical binding of rhizosphere soil, other physico-chemical effects such as cation binding and the production of iron/aluminium ligands will also be important.

PHYTOTOXICITY

There is a wide range of phytotoxic metabolites which can potentially be produced by soil micro-organisms (Lynch, 1976). The major focus of attention has again been saprophytes which utilise plant residues as substrates. For example acetic acid can be produced in phytotoxic concentrations during the decomposition of straw (Lynch, 1977; Tang & Waiss; 1978) and this will have a direct influence on roots when they grow close to residues. Presumably there is scope for such metabolites to be produced from root-derived carbon, especially when a suitable inoculum is provided by the proximity of the plant residue. A critical factor determining this will be whether the necessary reducing conditions develop around the root. This is an area which has been little studied and could be profitable. It is important to recognize such harmful as well as beneficial effects of rhizosphere micro-organisms. It might for example be possible to manipulate the system by applying simple chemicals, such as calcium peroxide (Sladdin & Lynch, 1983) to poise the redox potential of the rhizosphere above the levels which support fermentative metabolism.

ROOT INVASION

Whereas the decomposition of plant residues commonly has a negative effect on plant growth as a consequence of metabolite formation rather than a direct colonisation of the roots by the micro-organisms *per se* (Chapman & Lynch, 1984), there are some situations where bacteria can colonize the endorhizosphere and produce inhibitory effects on the growth of winter wheat (Elliott &

Lynch, 1984). This effect depends on the wheat cultivar, older varieties appearing more resistant. However, the effect, in common with some other rhizosphere effects is unstable, being non-repeatable in successive experiments. The suspicion is that the effect is mediated by a plasmid (J. Fredrickson, unpublished) but this awaits conclusive proof. This has wide implications for the whole approach to rhizosphere manipulation. Is it that the original 'azotobacterin' or the plant-growth promoting rhizobacteria (Kloepper *et al.*, 1980) are variable in their effects for this reason? Clearly this is an area worthy of investigation because if again the properties of the effect can be characterised, it should be possible to stabilise them.

CONCLUSION

The early studies of the rhizosphere by F.E. Chase, A.G. Lochhead, H. Katznelson, M.I. Timonin and others at the Department of Agriculture Canada, between 1938–47 provided a sound foundation in the characterisation of bacteria in the rhizosphere. In the more recent studies outlined in this review the population dynamics and function of rhizosphere organisms has been addressed. These studies are leading to the 'third era of rhizosphere studies' where manipulation of the rhizosphere, either by inoculation of natural or genetically-manipulated organisms and/or by modification of soil management practices, can be attempted. In this approach it is vital that population and function studies are continued. A major area which must be addressed is the study of the physiology of rhizosphere organisms. This will hopefully avoid further 'false-starts' in the use of rhizosphere inoculants and bring scientific and commercial credibility to the integration of an 'agrobiological sector' of the agrochemical industry (Lynch, 1983).

References

Barber, D.A., Bowen, G.D. & Rovira, A.D. (1976). Effects of micro-organisms on the absorption and distribution of phosphate in barley. *Australian Journal of Plant Physiology*, **3**, 801–808.

Barber, D.A. & Lee, R.B. (1974). The effect of micro-organisms on the absorption of manganese by plants. *New Phytologist*, **73**, 47–106.

Barber, D.A. & Martin, J.K. (1976). The release of organic substances by cereal roots in soil. *New Phytologist*, **76**, 69–80.

Barber, D.A. & Lynch, J.M. (1977). Microbial growth in the rhizosphere. *Soil Biology & Biochemistry*, **9**, 305–308.

Bennett, R.A. & Lynch, J.M. (1981a). Colonisation potential of rhizosphere bacteria. *Current Microbiology*, **6**, 137–138.

Bennett, R.A. & Lynch, J.M. (1981b). Bacterial growth and development in the rhizosphere of gnotobiotic cereal plants. *Journal of General Microbiology*, **125**, 95–102.

Brian, P.W. (1957). The ecological significance of antibiotic production. In *Microbial Ecology* (R.E.O. Williams & C.C. Spicer, eds.), pp. 168–188. Cambridge University Press; Cambridge.

Brown, M.E. (1974). Seed and root bacterization. *Annual Review of Phytopathology*, **12**, 181–197.

Chapman, S.J. & Lynch, J.M. (1984). The relative role of micro-organisms and their metabolites in the phytotoxicity of decomposing plant residues. *Plant & Soil*, **74**, 457–459.

Clarholm, M. (1983). Dynamics of soil bacteria in relation to plants, protozoa and inorganic nitrogen. Swedish University of Agricultural Sciences, Department of Microbiology, Uppsala, Report 17.

Darbyshire, J.F. & Greaves, M.P. (1973). Bacteria and protozoa in the rhizosphere. *Pesticide Science*, **4**, 349–360.

Dennis, C. & Webster, J. (1971). Antagonistic properties of species of *Trichoderma*. II. Production of volatile antibiotics. *Transactions of the British Mycological Society*, **57**, 41–48.

Dobereiner, J. & Day, J.M. (1976). Associative symbioses in tropical grasses. Characterization of micro-organisms and dinitrogen-fixing sites. In *Proceedings of the First International Symposium on Nitrogen Fixation* (W.E. Newton & C.J. Nyman, eds.), Vol. 2, pp. 518–538. Washington State University; Pullman.

Ebben, M.H. & Spencer, D.M. (1978). The use of antagonistic organisms for the control of black root rot of cucumber, *Phomopsis sclerotioides*. *Annals of Applied Biology*, **89**, 103–106.

Elad, Y., Chet, I. & Henis, Y. (1982). Degradation of plant pathogenic fungi by *Trichoderma harzianum*. *Canadian Journal of Microbiology*, **28**, 719–725.

Elad, Y., Chet, I., Boyle, P. & Henis, Y. (1983). Parasitism of *Trichoderma* spp. on *Rhizoctania solani* and *Sclerotium rolfsii*–scanning electron microscopy and fluorescence microscopy. *Phytopathology*, **73**, 85–88.

Elliott, L.F. & Lynch, J.M. (1984). Pseudomonads as a factor in the growth of winter wheat (*Triticum aestivum* L.). *Soil Biology & Biochemistry*, **16**, 69–71.

Elliott, L.F. & Lynch, J.M. (1985). Plant growth-inhibitory pseudomonads colonizing winter wheat (*Triticum aestivum* L.) roots. *Plant & Soil*. In press.

Focht, D.D. & Verstraete, W. (1977). Biochemical ecology of nitrification and denitrification. *Advances in Microbial Ecology*, **1**, 135–214.

Gilmour, J.T. & Gilmour, C.M. (1978). Nitrogenase activity of rice plant roots. *Soil Biology & Biochemistry*, **10**, 261–264.

Lynch, J.M. (1976). Products of soil micro-organisms in relation to plant growth. *CRC Critical Reviews in Microbiology*, **5**, 67–107.

Lynch, J.M. (1977). Phytotoxicity of acetic acid produced in the anaerobic decomposition of wheat straw. *Journal of Applied Bacteriology*, **42**, 81–87.

Lynch, J.M. (1983). *Soil Biotechnology. Microbiological Factors in Crop Productivity*. Blackwell Scientific Publications; Oxford.

Lynch, J.M. & Bragg, E. (1985). Micro-organisms and soil aggregate stability. *Advances in Soil Sciences*, **2**, in press.

Lynch, J.M. & White, N. (1977). Effects of some non-pathogenic micro-organisms on the growth of gnotobiotic barley plants. *Plant & Soil*, **47**, 161–170.

Mishustin, E.N. (1970). The importance of non-symbiotic nitrogen-fixing micro-organisms in agriculture. *Plant & Soil*, **32**, 545–554.

Newman, E.I. (1985). The rhizosphere: carbon sources and microbial populations. In *Ecological Interactions in the Soil Environment* (A.H. Fitter, ed.), in press. Blackwell Scientific Publications; Oxford.

Old, K.M. & Nicolson, T.H. (1978). The root cortex as part of a microbial consortium. In *Microbial Ecology* (M.W. Lortit & J.A.R. Miles, eds.), pp. 291–294. Springer-Verlag; Berlin.

Powell, P.E., Cline, C.R., Reid, C.P.P. & Szaniszlo, P.V. (1980). Occurrence of hydroxamate siderophore iron chelators in soils. *Nature*, **287**, 833–834.

Powell, P.E., Szansizlo, P.J., Cline, G.R. & Reid, C.P.P. (1982). Hydroxamate siderophore in the iron nutrition of plants. *Journal of Plant Nutrition*, **5**, 653–673.

Reid, J.B. & Goss, M.J. (1981). Effect of living roots of different plant species on the aggregate stability of two arable soils. *Journal of Soil Science*, **32**, 521–541.

Sladdin, M. & Lynch, J.M. (1983). Effect of calcium peroxide, lime and other seed dressings on winter wheat establishment under wet conditions. *Crop Protection*, **2**, 113–119.

Szaniszlo, P.J., Powell, P.E., Reid, C.P.P. & Cline, G.R. (1981). Production of hydroxamate siderophore iron chelators by ectomycorrhizal fungi. *Mycologia*, **73**, 1158–1174.

Tang, C.S. & Waiss, A.C. (1978). Short-chain fatty acids as growth inhibitors in decomposing wheat straw. *Journal of Chemical Ecology*, **4**, 225–232.

Whipps, J.M. (1984). Environmental factors affecting the loss of carbon from the roots of wheat and barley seedlings. *Journal of Experimental Botany*, **35**, 767–773.

Whipps, J.M. & Lynch, J.M. (1984). Substrate flow and utilization in the rhizosphere of cereals. *New Phytologist*, **95**, 605–623.

Whipps, J.M. & Lynch, J.M. (1985). The influence of the rhizosphere on crop productivity. *Advances in Microbial Ecology*, in press.

Nitrogen Fixation in a Sustainable Agriculture

J.I. Sprent

Department of Biological Sciences, University of Dundee, Dundee, Scotland

INTRODUCTION

Because it has been the subject of numerous books and symposia over recent years, this paper will assume readers are familiar with the general principles of nitrogen fixation (if not, see Sprent, 1979 and Postgate, 1982). In the context of *sustainable* agriculture, nitrogen fixation should be considered in relation to the biotic and abiotic environments; to this end the present paper will address four questions:

1. When and where is nitrogen fixation necessary?
2. Which types of nitrogen fixing organisms are likely to be of most use in sustainable agriculture?
3. What are the implications of nitrogen fixation for overall plant nutrient relations?
4. What biotic factors interact with nitrogen fixation?

The effects of environment on nitrogen fixation will not be covered as they have been reviewed elsewhere (Sprent *et al.*, 1983). The emphasis will be on general concepts rather than specific examples, since the latter will be discussed in many of the contributed papers. The coverage of the literature has, of necessity been rather selective.

WHEN AND WHERE IS NITROGEN FIXATION NECESSARY?

In a discussion on the evolution of nitrogen fixation, Sprent & Raven (1985) recently concluded that selective pressures in favour of nitrogen fixation required that combined nitrogen was the *major* limiting factor in the environment. Under circumstances where other nutrients (including water) were in short supply alternative strategies were adopted. The reasons behind this argument took into account not only the frequently discussed high energy costs of nitrogen fixation, but also costs in terms of other nutrients and the fact that the nitrogenase reaction is much *slower* than alternatives such as nitrate reduction. Further, in symbiotic systems such as legumes, nitrogen fixation requires a considerable *investment*, in that nodules must be formed before nitrogen fixation can begin. These arguments were developed further by Sprent (1985) in considering nitrogen fixing plants for arid environments.

Quantitatively, it was argued by Sprent & Raven (1985) that the nitrogen used to make a root nodule could, under conditions when nutrients generally are in short supply, be better used to make a greater volume of root tissue (root tissue being lower in N content than nodule tissue) which could not only scavenge soil for combined nitrogen, but also other nutrients and water. Even "typical" roots are not always the best way of obtaining nutrients from the soil. Under conditions of nutrient stress a larger absorptive surface for the *same* dry matter input can be achieved by the production of more root hairs, proteoid roots or the development of mycorrhizal symbioses. The latter are considered elsewhere in this volume and have also been recently reviewed in the context of nitrogen fixation (Barea & Azcon-Aguilar, 1983). Lamont (1982) has considered general strategies for obtaining nutrients in plants of mediterranean-type habitats. He has also (Lamont, 1983) made quantitative observations on the increase in surface area achieved when making proteoid roots rather than normal roots. In *Leucodendron laureolum* the ratio of surface to weight (mm^2 mg^{-1}) was 4658 and 295 for proteoid and normal roots respectively and it was calculated that proteoid roots could explore over 32 times the soil volume as a similar weight of normal root tissue. Similarly, surface area of root hairs may be greatly increased when phosphate or nitrate supplies are low, maximum increases of ninefold being found for spinach (Foehse & Jungk, 1983). For a theoretical consideration of this aspect see Robinson & Rorison (1983).

The above strategies for increasing nutrient uptake may also be found in nitrogen fixing higher plants, both legume and non-legume. For example, the Western Australian legume genus *Viminaria* and the actinorhizal genus *Casuarina* may produce any or all of nodules, mycorrhizas or proteoid roots (Lamont, 1972; National Research Council, 1984). Root growth of soybeans is greatly affected by soil P and K (Hallmark & Barber, 1984).

Thus, when considering nitrogen fixation, one should not lose sight of the other nutrient requirements of the plant. Some examples specific to nitrogen fixation will be considered in the third section.

In global terms, nitrogen fixation (biological or abiological) is necessary for two reasons. First to balance denitrification and second to supply the net increase (if any) in world biomass. However, in local terms there may be considerable need for nitrogen fixation to balance losses by removal of the crop (including grazing by herbivores) and by burning. The latter, unless necessary for fuel purposes should only be used as a last resort and even then residues returned to the land. Wherever possible green or dry crop residues should be ploughed in or used for silage, to minimise losses of essential nutrients other than nitrogen (see the third section). A case history is presented in the following paper (pp.167–189).

WHAT TYPES OF NITROGEN FIXING ORGANISM ARE LIKELY TO BE OF MOST USE IN SUSTAINABLE AGRICULTURE?

The answers to this question will vary greatly with geographical location. The major categories of nitrogen fixing organism will be considered in turn.

Free living heterotrophic bacteria, e.g. *Azotobacter*

These, like other heterotrophic bacteria need a source of carbon from the soil. For reasons discussed in the previous section, nitrogen fixing species are only likely to be at a competitive advantage when soil N is low. Figure 1 gives data from a wide range of soil horizon types, showing the relation between soil organic matter and the C:N ratio of that matter. Generally, soils low in organic matter have a lower C:N ratio than those higher in organic matter. It is in the latter that nitrogen fixing organisms are most likely to be an advantage, although problems such as low pH and waterlogging may keep microbial activity low. However, these general arguments may be put to practical use in the amendment of soils with carbon-rich residues, combined with inoculation with a suitable nitrogen fixing organism if necessary. For example, Patriquin (1982) obtained high rates of N fixation in sugar cane trash in the West Indies. Use of straw as a soil amendment, inoculated with nitrogen fixing organisms is being tried in various parts of the world and has great potential. If, as is generally agreed trace element cations are essentially phloem immobile and therefore not transported out of senescing tissue (see, for example, Kirkby & Pilbeam, 1984), the sheathing leaf bases of grass/cereal residues may be particularly useful in meeting the special needs of nitrogen fixing systems (see next section).

In considering free living heterotrophs as well as all other categories of nitrogen fixing organisms, one should not lose sight of other attributes which they may possess favouring (or inhibiting) plant growth. In the cases of *Azotobacter* and *Azospirillum* (see below), production of growth substances may be, arguably, of equal importance to nitrogen fixation (see, for example, Brown, 1982).

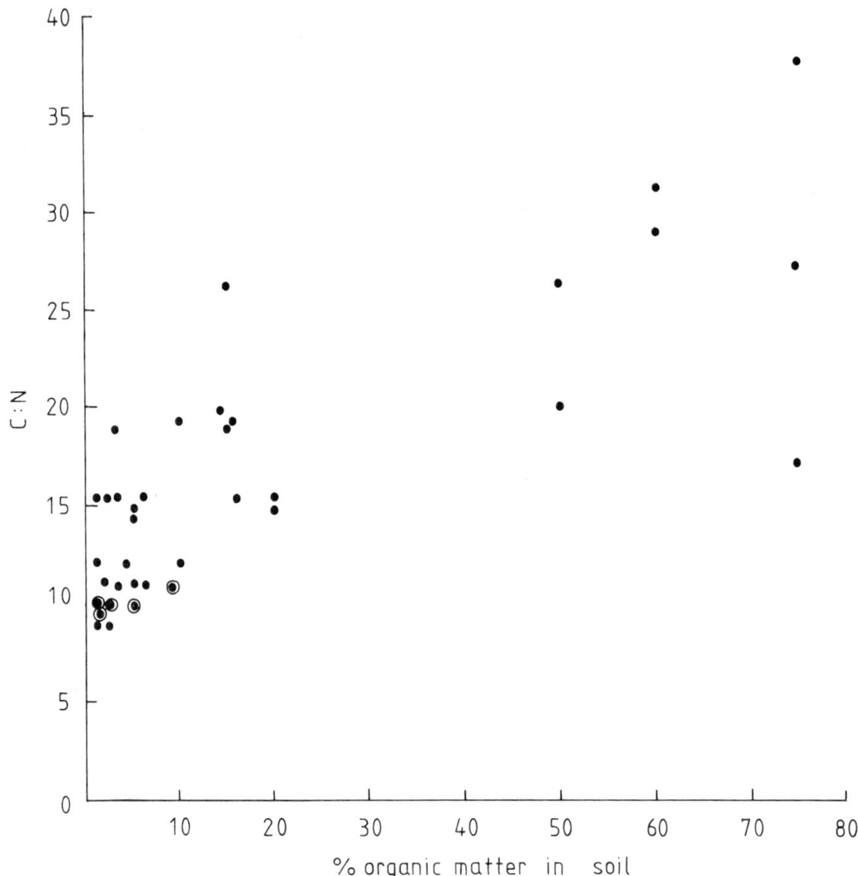

FIGURE 1   The relation between % organic matter in various soils and the C:N ratio of that organic matter. Data plotted from Fitzpatrick (1983). Where a range of values was given, a mean is plotted, where values ⟨ or ⟩ given, the limiting value. Points surrounded by a circle represent more than one (up to eight) values.

Photosynthetic nitrogen fixing prokaryotes

Agriculturally, the cyanobacteria, either free living or in symbiosis, are the most significant group. Although some are very drought tolerant (see discussion in Sprent, 1985), they can only fix nitrogen in the hydrated state and their greatest significance is likely to be in partially aquatic systems, notably the production of wetland rice. Both free living cyanobacteria and, more frequently, *Azolla* spp. have played a pivotal role in sustainable agriculture in the Far East for many years (Moore, 1969) and will not be further considered here.

Associations with plant roots

For a description of the bacteria involved, see Postgate (1982). Much work has been carried out on these in recent years, but the extent of their significance remains to be elucidated. Undoubtedly, many pure grass pastures in the tropics appear to grow well for many years in low N soils and without N fertilizer. In this low input-low output system, nitrogen fixing bacteria associated with roots (for example *Azospirillum lipoferum*) may play a significant role (Patriquin *et al.*, 1983).

Actinorhizas

With a few exceptions, these associations between non-leguminous dicotyledons and the filamentous bacterium *Frankia*, are of more significance to forestry than agriculture. A few, such as *Hippophae rhamnoides* produce edible and marketable berries and the genus *Casuarina* is important as a shade plant.

Legumes

These are by far the most significant group of nitrogen fixing plants in agriculture and most of the succeeding discussion will centre on them. Pastures composed of legumes and grasses are probably the best known examples of sustainable agriculture and these will not be considered further here. The use of grain and shrub legumes is less widely understood in this context.

Most studies of grain legumes have been on soybean, peas and *Phaseolus* and *Vicia* beans, generally under intensive agriculture. Unfortunately, in breeding for increased yield and nutritional quality, the natural defences of the legume against pests have been breached. The high level of N in legume seed, which may be a necessary requisite for investment in root nodules (see Sprent & Raven, 1985) also makes them ideal as food for pests. It is difficult to see a way out of this dilemma without using some form of pest control, preferably biological.

The advantages and exploitation of shrubby legumes are considered in detail elsewhere (Sprent, 1985 and other papers in the same Proceedings). They have great potential, particularly in harsh environments because they help to combat erosion and have multiple uses. By regular cropping (analogous to grazing, or cutting pastures), nitrogen fixation may be optimised.

WHAT ARE THE IMPLICATIONS OF NITROGEN FIXATION FOR OVERALL PLANT NUTRIENT RELATIONS?

Soil pH

Nitrogen nutrition has a major effect upon pH around plant roots. Plants assimilating ammonium give a considerable net acidification, those fixing nitrogen a lesser net acidification and those assimilating nitrate either a rise in pH, no change or a slight net acidification depending on the cation uptake proceeding at the time. For details of the arguments behind these statements, see Raven & Smith (1976); Sprent (1984) has summarised the processes in the context of nitrogen fixation. Thus nitrogen nutrition may profoundly affect the pH of poorly buffered soils (Jarvis & Robson, 1983). Even when soils are well buffered, local effects (which can be seen *in situ* using pH indicator dyes, see Marschner & Romheld, 1983) could be important. In particular, changes in pH are likely to affect the availability of certain macro- and micronutrients. Low pH results in solubilisation of rock phosphate and nitrogen fixing plants may be able to obtain more phosphate than nitrate assimilating plants, even without the undoubted beneficial effects of VA mycorrhizas, considered elsewhere in this volume (Bekele *et al.*, 1983; Barea & Azcon-Aguilar, 1983). In addition, soil phosphatases and other enzymes are affected by changes in soil pH (Frankenberger & Johanson, 1982).

Trace element availability is affected in various ways by changes in soil pH. Molybdenum, which is required both for nitrogen fixation and nitrate reduction (see Sprent *et al.*, 1983; Shah *et al.*, 1984 and Table 1) becomes unavailable at low pH whereas iron (and probably nickel) become more available. Iron is an essential nutrient for all plants, but nitrogen fixing legumes have a particularly high requirement for the synthesis of leghaemoglobin (the role of this pigment has been reviewed by Appleby, 1984) as well as nitrogenase (see below).

Micronutrient relationships

Table 1 summarises the particular requirements of nitrogen fixing legumes; those listed under rhizobia may be common to all nitrogen fixing prokaryotes. However, it is important to note that the host requirement may vary amongst legumes. For example, molybdenum, as well as being a constituent of nitrogenase is also found in xanthine dehydrogenase. The latter enzyme may be necessary in small quantities in all plants, but is required in much larger amounts and is specifically formed in those legumes (e.g. *Glycine, Vigna, Phaseolus, Cajanus*) which export ureides as a product of nitrogen fixation (see Sprent, 1980; Thomas & Schrader, 1981).

TABLE 1

Trace element requirements specifically[1]
associated with nitrogen fixation and related
reactions in legumes

Element	*Rhizobium*	host
Fe	N_2ase	haemoglobin
Mo	N_2ase	xanthine dehydrogenase[2]
Ni	uptake H_2ase[3]	urease[2]

[1]Co and Cu are also needed but their exact role is not clear.
[2]In some legume tribes only
[3]Not present in all strains

Nickel is not generally recognised as an essential element for plants. Recently it has been found to be a constituent of plant (as of animal) urease (Klucas *et al.*, 1983). Many legumes contain high levels of urease—indeed the major commercial source of urease is seed of the jack bean (*Canavalia ensiformis*), also a ureide exporting legume. Urease is particularly significant in these since the likely pathway from ureides to amides and amino acids is *via* urea. Nickel is also constitutive for uptake hydrogenase in bacteria, including nitrogen fixing species such as rhizobia (e.g. Stults *et al.*, 1984). The possible significance of uptake hydrogenase is considered below.

Cobalt (see references in Sprent *et al.*, 1983) and copper (Snowball *et al.*, 1980) also appear to be involved specifically with the nitrogen fixing process in legumes, but since the exact location of their involvement is not known, we shall not consider them further here.

It can be seen from Table 1 that iron, molybdenum and nickel are required by *both* symbionts in legume nodules. This could imply that they were both available in soil at the time when legume nodules evolved, generally considered to be in the upper Cretaceous (see discussion in Sprent & Raven, 1985). It is interesting therefore that a recent paper (Smit & Kyte, 1984) suggested that some rock components may have been high in nickel at about that time. The same components were low in silicon oxide: in present soils nickel is usually unavailable because it is precipitated as insoluble nickel silicates. This fact has encouraged people to think that it was an unlikely candidate for an essential plant element (Jameson, personal communication). This is now known not to be the case (see above). There has been little consideration of the evolution of current trace element requirements, but it seems likely that the relevant metalloenzymes may have evolved when the metals were in an available form. At the present time, there may be considerable difficulty in obtaining them from the soil. Iron has probably been the most widely studied. Microorganisms including nitrogen fixing species such as *Azotobacter* (Knosp *et al.*, 1984) may invest a considerable amount of carbon and energy making substances (called siderophores) which enable them to take up iron from soils. Such substances may persist in soils (Powell *et al.*, 1980) and be at particularly high concentrations in the rhizosphere (Reid *et al.*, 1984). They may assist plant uptake of iron, supplementing any mechanisms induced in plants by low iron availability, for example formation of epidermal transfer cells (Kramer *et al.*, 1980). It is possible, although no evidence is known to the author, that these siderophores could also aid uptake of nickel. Thus nitrogen fixing organisms, together with the general soil flora, may help to maintain soil nutrients in an available form.

WHAT BIOTIC FACTORS INTERACT WITH NITROGEN FIXATION?

We have seen above that microbiological production of siderophores may influence iron uptake by plants. In this section further examples of biotic effects will be considered.

Reactions involving hydrogen

It has been known for several years that, concomitantly with reducing nitrogen to ammonia, nitrogenase reduces protons to hydrogen gas. It now appears that the production of hydrogen is incvitable (see Simpson & Burris (1984) and references therein). However, most nitrogen fixing organisms have the genetic potential to recycle hydrogen using a Ni-containing uptake hydrogenase which is usually coupled to ATP generation (see references in Sprent, 1984). This hydrogen recycling may recover some of the energy lost in its production. However, many strains of rhizobia do not have an uptake hydrogenase and hydrogen gas may be released into the soil where it has been suggested to be a significant part of the global hydrogen budget (Conrad & Seiler, 1980). However, many soil organisms apart from those fixing nitrogen contain uptake hydrogenases and, depending on the content of these in the soil there may or may not be net hydrogen evolution (Hopmans *et al.*, 1983). Reactions involving hydrogen are thus likely to be reflected in the microbial flora and to be concerned in the redox state of the soil (see also Conrad *et al.*, 1983): nitrogen fixation plays a significant role in these relationships.

Allelopathy and other negative interactions

Allelopathy has been neglected as a factor involved in nitrogen fixation *per se*. Dawson & Seymour (1983) found that juglone from black walnut (*Juglans nigra*) inhibited growth of both *Frankia* and *Rhizobium*: they pointed out that such effects could be important in choice of nitrogen fixing trees for nurse crops. Allelopathic effects were also implicated in results from experiments using different planting configurations of *Alnus* and *Populus* (Dawson *et al.*, 1983). Spacing of nitrogen fixing pasture plants responds naturally to factors such as exploitation of available nutrients (see, for example, Turkington & Harper, 1979). These considerations should be taken into account in sustainable agriculture, especially with respect to mixed cropping systems.

Related to the topic of allelopathy is the recent report that seeds of some cultivars of arrowleaf clover contain thermostable, water soluble substance(s) toxic to *Rhizobium trifolii* (Materon & Weaver, 1984).

Sporadic reports of foliar chlorosis produced by rhizobitoxine, a substance synthesised by some strains of rhizobia, have appeared over the years, and the effects are rather specific to plant genotype (e.g. Devine *et al.*, 1983). More recently, leaf roll symptoms in pigeon pea were attributed to *Rhizobium* (Rao *et al.*, 1984). Whether either of these effects acts via plant growth substance imbalance is not known, but rhizobia are known to produce plant growth regulating substances (see references in Sprent, 1984). Rhizobitoxine may have a positive role to play in that it may protect soybean roots from charcoal root disease (caused by the fungus *Macrophomina phaseolina*, Chakraborty & Porkayastha, 1984). By affecting the general nutrient status of the plant, *Rhizobium* could have considerable interaction with plant disease. The general subjection of nutrient stress and plant disease has been reviewed recently, but little is known about the role of symbiosis (Graham, 1983).

Microbial interactions involving nitrogen fixing organisms

Two of these have already been considered: those involving hydrogen and the acquisition of iron. Interesting interactions have been proposed whereby microorganisms may enhance nodulation. This has been shown for *Pseudomonas cepacia* on nodulation of *Alnus rubra* by *Frankia* (Knowlton & Dawson, 1983), for *Azotobacter vinelandii* (Burns *et al.*, 1981) and *Azospirillum lipoferum* (Tilak *et al.*, 1981) on nodulation of various legumes. The effect of *Pseudomonas putida* in improving nodulation of *Phaseolus vulgaris* by *R. phaseoli* was thought to have been mediated *via* improved phosphate supply, since *P. putida* produces a phosphate solubilizing organic acid (Grimes & Mount, 1984).

A further effect of *Rhizobium* is that it may help inactive toxins produced by soil fungi, e.g. *Metorhizium* sp. (Habte & Barrion, 1984).

Nitrogen fixation in legumes in relation to plant diseases

Not surprisingly, when nitrogen fixing organisms bear pests or pathogens, the nitrogen fixing process is usually adversely affected. Conversely, the nitrogen fixing symbiosis may affect the plant disease. It has been suggested above that this could result from effects on plant nutrient status as well as direct effects. Some further points from a range of diseases will be considered here, using mainly recent publications since the earlier literature has been reviewed elsewhere (Bowen, 1978).

Plant viruses affect the structure, physiology and biology of both root nodules and host plant. The subject has been reviewed recently (Tu & Ford, 1984). Amongst the interesting points raised by these authors is that a virus affected plant may, *via* root exudates, selectively affect rhizobial strains. The significance of this observation for inoculation is clear. Nodules may have very high levels of virus and, apart from affecting all the standard measures of nodule activity (acetylene reduction, leghaemoglobin content), enhanced nodule nitrate reductase activity has been reported for red clover infected with white clover mosaic virus (Khadhair *et al.*, 1984).

Many interactions between plant pathogenic fungi and root nodules have been reported. Cases of plant protection by rhizobia have been cited in the previous section. Tu (1978) found that rhizobia actually colonised the tips of hyphae of *Phytophthora megasperma*, which causes root rot of soybeans and lucerne (alfalfa) and in this way could affect fungal growth. In lucerne, this pathogen is particularly significant at low winter temperatures and it is under such conditions that rhizobial protection may be most significant (Tu, 1980).

On the pest side, root nodules may be eaten by *Sitona* weevil larvae (see, for example, Bowen, 1978). Nematodes, apart from effects *via* host vigour, may specifically result in production of nodulation suppressing factors (Ko *et al.*, 1984). On the other hand, legumes produce many toxins to pests. One strategy specific to nitrogen fixing plants may occur in those legumes which export ureides. These substances are not utilised by insect herbivores and thus nitrogen fixing plants (compared with plants of the same genotype utilising combined nitrogen) may be at an advantage, because they provide a less nutritionally useful food (Wilson & Stinner, 1984).

GENERAL COMMENTS AND CONCLUSIONS

In the above account I have tried to indicate some of the factors which should be taken into account when considering nitrogen fixation in the context of sustainable agriculture. The interactions with soil, climatic and biotic factors are legion and will vary greatly from region to region. Obviously nitrogen fixing plants must play an integral role in sustainable agriculture, but before we can use them to most benefit we need to know a great deal more about the physiology of the different systems and their interaction with their environment, both biotic and abiotic. For example, until recently almost all trace element studies in legumes have been carried out on plants given combined nitrogen. Since legumes are likely to be the most important nitrogen fixing systems, I cannot do better than reiterate the advice of Tinker (1984) that more attention should be paid to the morphology and physiology of the host legume to complement present work on *Rhizobium* alone and add that this attention should be given under realistic (preferably field) conditions.

References

Appleby, C.A. (1984). Leghemoglobin and *Rhizobium* respiration. *Annual Review of Plant Physiology*, **35**, 443–478.

Barea, J.M. & Azcon-Aguilar, C. (1983). Mycorrhizas and their significance in nodulating nitrogen fixing plants. *Advances in Agronomy*, **36**, 1–54.

Bekele, T., Cino, B.C., Ehlert, R.A.I., van der Maas, A.A. & van Diest, A. (1983). An evaluation of plant-borne factors promoting the solubilization of alkaline rock phosphate. *Plant & Soil*, **75**, 361–378.

Bowen, G.D. (1978). Dysfunction and shortfalls in symbiotic responses. In *Plant Disease III* (J.G. Horsfall & E.B. Cowling, eds.), pp. 231–256. Academic Press; New York.

Brown, M.E. (1982). Nitrogen fixation by free living bacteria associated with plants—fact or fiction? In *Bacteria and Plants* (M.E. Rhodes-Roberts & F.A. Skinner, eds.), pp. 25–41. Academic Press; London.

Burns, T.A. Jr., Bishop, P.E. & Israel, D.W. (1981). Enhanced nodulation of leguminous plant roots by mixed cultures of *Azotobacter vinelandii* and *Rhizobium*. *Plant & Soil*, **62**, 399–412.

Charkraborty, V. & Porkayastha, R.P. (1984). Role of rhizobitoxine in protecting soybean roots from *Macrophomina phaseolina* infection. *Canadian Journal of Microbiology*, **30**, 285–289.

Conrad, R. & Seiler, W. (1980). Contribution of hydrogen production by biological nitrogen fixation to the global hydrogen budget. *Journal of Geophysical Research*, **85**, 5493–5498.

Conrad, R., Weber, M. & Seiler, W. (1983). Kinetics and electron transport of soil hydrogenases catalyzing the oxidation of atmospheric hydrogen. *Soil Biology & Biochemistry*, **15**, 167–173.

Dawson, J.O. & Seymour, P.E. (1983). Effects of juglone concentration on growth of *Frankia* Ar13 and *Rhizobium japonicum* strain 71 *in vitro*. *Journal of Chemical Ecology*, **9**, 1175–1183.

Dawson, J.O., Dzialowy, P.J., Gertner, G.Z. & Hansen, E.A. (1983). Changes in soil nitrogen concentration around *Alnus glutinosa* in a mixed, short-rotation plantation with hybrid *Populus*. *Canadian Journal of Forestry Research*, **13**, 572–576.

Devine, T.E., Kuykendall, L.D. & Breithaupt, B.H. (1983). Nodule-like structures induced on peanut by chlorosis producing strains of *Rhizobium* classified as *R. japonicum*. *Crop Science*, **23**, 394–397.

Fitzpatrick, E.A. (1983). *Soils*. Longman; London.

Foehse, D. & Jungk, A. (1983). Influence of phosphate and nitrate supply on root hair formation of rape, spinach and tomato plants. *Plant & Soil*, **74**, 359–368.

Frankenberger, W.T.Jr. & Johanson, J.B. (1982). Effects of pH on enzyme stability in soils. *Soil Biology & Biochemistry*, **14**, 433–437.

Graham, R.D. (1983). Effects of nutrient stress on susceptibility of plants to disease, with particular reference to trace elements. *Advances in Botanical Research*, **10**, 222–276.

Grimes, H.D. & Mount, M.S. (1984). Influence of *Pseudomonas putida* on nodulation of *Phaseolus vulgaris*. *Soil Biology & Biochemistry*, **16**, 27–30.

Habte, M. & Barrion, M. (1984). Interactions of *Rhizobium* sp. with toxin-producing fungus in culture medium in a tropical soil. *Applied Environmental Microbiology*, **47**, 1080–1083.

Hallmark, W.B. & Barber, S.A. (1984). Root growth and morphology, nutrient uptake, and nutrient status of early growth of soybeans as affected by soil P and K. *Agronomy Journal*, **76**, 209–212.

Hopmans, P., Chalk, P.M. & Douglas, L.A. (1983). Symbiotic N_2 fixation by legumes grown in pots II. Uptake of ^{15}N labelled NO_3^-, C_2H_2 reduction and H_2 evolution by *Trifolium subterraneum* L., *Medicago truncatula* Gaertn. and *Acacia dealbata* Link. *Plant & Soil*, **74**, 333–342.

Jarvis, S.C. & Robson, A.D. (1983). The effect of nitrogen nutrition of plants on the development of acidity in Western Australian soils I. Effects with subterranean clover from under leaching conditions. *Australian Journal of Agricultural Research*, **34**, 341–353.

Khadhair, A.H., Sinha, R.C. & Peterson, J.F. (1984). Effect of white clover mosaic virus infection on various processes relevant to symbiotic N_2 fixation in red clover. *Canadian Journal of Botany*, **62**, 38–42.

Kirkby, E.A. & Pilbeam, D.J. 1984. Calcium as a plant nutrient. *Plant Cell Environment*, **7**, 397–405.

Klucas, R.V., Hanus, F.J., Russell, S.A. & Evans, H.J. (1983). Nickel: a micronutrient element for hydrogen-dependent growth of *Rhizobium japonicum* and for expression of urease activity in soybean plants. *Proceedings of the National Academy of Sciences*, **80**, 2253–2257.

Knosp, O., von Tigerstrom, M. & Page, W.J. (1984). Siderophore-mediated uptake of iron in *Azotobacter vinelandii*. *Journal of Bacteriology*, **159**, 341–347.

Knowlton, S. & Dawson, J.O. (1983). Effects of *Pseudomonas cepacia* and cultural factors on the nodulation of *Alnus rubra* roots by *Frankia*. *Canadian Journal of Botany*, **61**, 2877–2882.

Ko, M.P., Barker, K.R. & Huang, J-S. (1984). Nodulation of soybeans as affected by half-root infection with *Heterodera glycines*. *Journal of Nematology*, **16**, 97–105.

Kramer, D., Romheld, V., Landsberg, E. & Marschner, H. (1980). Induction of transfer-cell formation by iron deficiency in the root epidermis of *Helianthus annuus* L. *Planta*, **147**, 335–339.

Lamont, B. (1972). "Proteoid" roots in the legume *Viminaria juncea*. *Search*, **3**, 91–92.

Lamont, B. (1982). Mechanisms for enhancing nutrient uptake in plants with particular reference to Mediterranean South Africa and Western Australia. *Botanical Review*, **48**, 597–689.

Lamont, B. (1983). Root hair dimensions and surface/volume/weight ratios of roots with the aid of scanning electron microscopy. *Plant & Soil*, **74**, 149–152.

Marschner, H. & Romheld, V. (1983). *In vivo* measurement of root-induced pH changes at the soil-root interface: effect of plant species and nitrogen source. *Zeitschrift fur Pflanzenphysiologie*, **111**, 241–251.

Materon, L.A. & Weaver, R.W. (1984). Toxicity of arrowleaf clover seed to *Rhizobium trifolii*. *Agronomy Journal*, **76**, 471–473.

Moore, A.W. (1969). *Azolla*: biology and agronomic significance. *Botanical Review*, **35**, 17–34.

National Research Council. (1984) Casuarinas: nitrogen fixing trees for adverse sites. National Academy of Sciences; Washington, D.C.

Patriquin, D.G. (1982). Nitrogen fixation in sugar cane litter. *Biological Agriculture & Horticulture*, **1**, 39–64.

Patriquin, D.G., Dobereiner, J. & Jain, D.K. (1983). Sites and processes of association between diazotrophs and grasses. *Canadian Journal of Microbiology*, **29**, 900–915.

Postgate, J.R. (1983). *Fundamentals of nitrogen fixation*. Cambridge University Press; Cambridge.

Powell, P.E., Cline, G.R., Reid, C.P.P. & Szaniszlo, P.J. (1980). Occurrence of hydroxamate siderophore iron chelators in soils. *Nature*, **287**, 833–834.

Rao, J.V.D.K.K., Dart, P.J. and Kiran, M.U. (1984). *Rhizobium* induced leaf roll in pigeonpea [*Cajanus cajan* (L.) Millsp]. *Soil Biology & Biochemistry*, **16**, 89–91.

Raven, J.A. & Smith, F.A. (1976). Nitrogen assimilation and transport in relation to intracellular pH regulation. *New Phytologist*, **76**, 425–431.

Reid, R.K., Reid, C.P.P., Powell, P.E. & Szaniszlo, P.J. (1984). Comparison of siderophore concentrations in aqueous extracts of rhizosphere and adjacent bulk soils. *Pedobiologia*, **26**, 263–266.

Robinson, D. & Rorison, I.H. (1983). Relationships between root morphology and nitrogen availability in a recent theoretical model describing nitrogen uptake from the soil. *Plant Cell Environment*, **6**, 641–647.

Shah, V.K., Ugalde, R.A., Imperial, J. & Brill, W.J. (1984). Molybdenum in nitrogenase. *Annual Review of Biochemistry*, **53**, 231–257.

Simpson, F.B. & Burris, R.H. (1984). A nitrogen pressure of 50 atmospheres does not prevent evolution of hydrogen by nitrogenase. *Science*, **224**, 1095–1095.

Smit, J. & Kyte, F.T. (1984). Siderophile-rich magnetic spheroids from the Cretaceous-Tertiary boundary in Umbria, Italy. *Nature*, **310**, 403–405.

Snowball, K., Robson, A.D. & Loneragan, J.F. (1980). The effect of copper on nitrogen fixation in subterranean clover (*Trifolium subterraneum*). *New Phytologist*, **85**, 63–72.

Sprent, J.I. (1979). *The biology in nitrogen fixing organisms*. McGraw-Hill; London.

Sprent, J.I. (1980). Root nodule anatomy, type of export product and evolutionary origin in some Leguminose. *Plant Cell Environment*, **3**, 35–43.

Sprent, J.I. (1984). Nitrogen fixation in arid environments. In *Plants for Arid Lands* (G.E. Wickens, J.R. Goodin & D.V. Field, eds.), pp.215–229. George Allen & Unwin; London.

Sprent, J.I. (1985). Nitrogen fixation. In *Proceedings of the Kew International Conference on Economic Plants for Arid Lands* (G. Wickens, ed.). In press.

Sprent, J.I., Minchin, F.R. & Thomas, R.J. (1983). Environmental effects on the physiology of nodulation and nitrogen fixation. In *Temperate Legumes* (D.G. Jones & D.R. Davies, eds.), pp.269–318. Pitman; London.

Sprent, J.I. & Raven, J.A. (1985). Evolution of nitrogen fixing symbioses. *Proceedings of the Royal Society of Edinburgh*, **85B**, 215–237.

Stults, L.W., O'Hara, E.B. & Maier, J. (1984). Nickel as a component of hydrogenase in *Rhizobium japonicum*. *Journal of Bacteriology*, **159**, 153–158.

Thomas, R.J. & Schrader, L.E. (1981). Ureide metabolism in higher plants. *Phytochemistry*, **20**, 361–371.

Tilaj, K.V.B.R., Singh,C.S. & Rana, J.P.S. (1981). Effects of combined inoculation of *Azospirillum brasilense* with *Rhizobium trifolii, Rhizobium meliloti* and *Rhizobium* sp (cowpea miscellany) on nodulation, and yield of clover (*Trifolium repens*), lucerne (*Medicago sativa*) and chick pea (*Cicer arietinum*). *Zentralblatt für Bakteriologie, Parasitenkunde, Infektsionskrankheiten und Hygiene, Abt. II*, **136**, 117–120.

Tinker, P.B. (1984). The role of microorganisms in mediating and facilitating the uptake of plant nutrients from soil. *Plant & Soil*, **76**, 77–91.

Tu, J.C. (1978). Protection of soybean from severe *Phytophthora* root rot by *Rhizobium*. *Physiological Plant Pathology*, **12**, 233–240.

Tu, J.C. (1980). Incidence of root rot and overwintering of alfalfa as influenced by rhizobia. *Phytopathologische Zeitschrift*, **97**, 97–108.

Tu, J.C. & Ford, R.E. (1984). Plant virus interaction in nitrogen-fixing nodules. *Zeitschrift für Pflanzenkrankheiten und Pflanzenschutz*, **91**, 200–212.

Turkington, R. & Harper, J.L. (1979). The growth, distribution and neighbour relationships of *Trifolium repens* in a permanent pasture. I. Ordination, pattern and contact. *Journal of Ecology*, **67**, 201–218.

Wilson, K.G. & Stinner, R.E. (1984). A potential influence of rhizobium activity on the availability of nitrogen to legume herbivores. *Oecologia*, **61**, 337–345.

Biological Husbandry and the "Nitrogen Problem"

David G. Patriquin

Biology Department, Dalhousie University, Halifax, Nova Scotia, Canada, B3H 4J1

INTRODUCTION

There are few agricultural systems today that operate without substantial inputs of nitrogen fertilizer. Because of the high cost of these inputs, and increasing concern over aquatic, groundwater and atmospheric pollution by N fertilizer (Stewart & Rosswall, 1982) considerable research is being conducted with the goal of making plants "self-sufficient" in nitrogen (Earl & Ausubel, 1983). However, a decade of intensive research into biological N_2 fixation has failed to bring about a substantial reduction in the use of nitrogen fertilizer, and there is little prospect that it will do so in the near future. We need to seriously consider, therefore, whether it is possible to practice agriculture with greatly reduced inputs of N fertilizer by using existing N_2-fixing resources in conjunction with recycling; in other words, to consider to what extent biological husbandry could reduce the requirements for nitrogen fertilizer.

Over the past five years I have had an opportunity to study the practical and theoretical aspects of this question. This opportunity resulted from meeting a farmer who had stopped using fertilizer and pesticides and was attempting to meet the N needs of his cereals by use of legumes and manure produced on the farm. In this paper I will describe what we have learned, factually and conceptually, about the "N problem" from our studies. By "we" I refer to Mr. Basil Aldhouse (the farmer), myself, and to a succession of Honors Biology students (David Burton, Nick Hill, Gillian Allan, Mary Bishop and Danica Baines) each of whom contributed in his or her own unique way to an enlarged vision of "the farm".

REGIONAL SETTING

The farm is located in the Annapolis Valley of Nova Scotia (Canada), a cool (120 frost free days), humid (114cm rain annually), temperate region. It consists of approximately 35 ha of field crops, 2 ha of pasture and garden, and 35 ha in woodland and water. The fields are level to rolling, contain 2.8 to 7.1% organic matter, vary in texture from loamy sand to sandy clay loam and are classified as being in categories 3 and 4 with moderately severe to severe limitations. Approximately 2000 laying hens are maintained in a traditional floor operation with deep litter. Mr. Aldhouse attempts to be self sufficient in feed; this number of birds is roughly that which he found could be supported by grain production on 30 ha of land managed conventionally. In 1975, he achieved the highest oat yield (98 bushels/acre) in a provincial competition. Out of concern over rising requirements for inputs and the large amounts of toxic materials he was using, he decided in 1976 to stop using fertilizer and pesticides and to see how the farm could manage on its own resources. His cereal yields promptly fell by about 50%. Rather than reduce the size of his flock, he purchased grains to make up the difference, and explored ways to increase his yields. From 1980 to 1983, yields of faba beans, winter wheat and oats averaged about 25% below those cited as normal for this region with full fertilization (2.7 to 3.1 tonnes/ha). In spite of those reductions, the farm has remained profitable; the cost opf purchasing grain to make up for the shortfalls in production is equal to or less than the cost he would otherwise have spent on fertilizer and pesticides. Oat yields improved substantially in 1984.

USING A GRAIN LEGUME TO SUPPLY N TO OTHER CROPS

We didn't set out to study Mr. Aldhouse's farm. We were interested in the faba bean (*Vicia faba* minor), and Mr. Aldhouse was one of the few farmers growing this plant locally.

The faba bean is a grain legume, grown traditionally in China, Europe, the Middle East, North Africa and Peru. Its use in Europe declined dramatically in the mid 20th century but there is now renewed interest in the crop (Thompson & Taylor, 1982). It was first grown commercially in Canada in 1967, by Robyn Warren, an Englishman farming the dykelands of Nova Scotia. Several other farmers, including Mr. Aldhouse, took it up in the next few years, and subsequently it was introduced to western Canada (Evans *et al.*, 1972).

As faba beans in western Canada had been reported to respond to inoculation (Candlish & Clark, 1975), we thought we might be able to increase N fixation on local farms by inoculating crops with "superior" *Rhizobium* strains. A preliminary survey revealed, however, that there was no need for inoculants. Plants were well nodulated and exhibited high nitrogenase activity (Patriquin & Burton, 1982). Further, the farmers were less interested in increased N_2 fixation, than they were in learning how to use existing N fixation. Mr. Warren showed me several examples of how maize and cereals exhibited better growth where they followed faba beans in a crop rotation. He assumed this was due to N_2 fixation and wanted to know how much he could allow for it in fertilizer applications. We supposed that you could reduce the N applications to subsequent crops by roughly the difference between the nitrogen fixed and the nitrogen removed at harvest.

However, a N budget for Mr. Warren's beans suggested that the bean crop was withdrawing substantially more N from the soil than it was putting in through N_2 fixation (Fig. 1). That particular site had an unusually high yield, but studies at sites of lower yields also indicated near zero or negative N balances (Patriquin *et al.*, 1981). Similarly, negative N balances have been found for soybean (Johnson *et al.*, 1975) and even some forage legumes when the latter are not consumed in the field (Rice, 1980).

Interestingly, it was apparently well known in the pre-chemical era that continuous culture of grain legumes leads to rapid decline in soil N (Harmsen & van Schreven, 1955).

How, then, does the faba bean benefit subsequent crops? Possibly by bringing up nutrients from deep horizons via their well developed tap roots, or by putting N in a highly available form, i.e. in high N residues (Fig. 1).

In any case, the negative or near zero N balances mean that N_2 fixed by this grain legume cannot reduce the net N requirement of other crops grown in rotation with it unless some of the legume-N removed at harvest is recycled, i.e. as manure.

RECYCLING GRAIN LEGUME N

That was in effect how Mr. Aldhouse was attempting to use N_2 fixed by the faba beans. He had beans on 1/3 of his land, fed the grains to his hens, and applied the hen manure to the cereal fields. Yet the cereals still suffered from a severe shortage of N. Why?

Figure 2 illustrates the major flows and reservoirs of N on the Aldhouse farm in 1979 (Patriquin *et al.*, 1981). For the moment, there are three points to be noted:

(1) There were roughly 5000 Kg N cycling around the farm and only 400 Kg being exported as eggs. It can immediately be appreciated that the greatest inefficiencies in conventional egg-producing systems result from separating the sites of N_2 fixation (e.g. in soybean) and of manure production from the sites of cereal production.

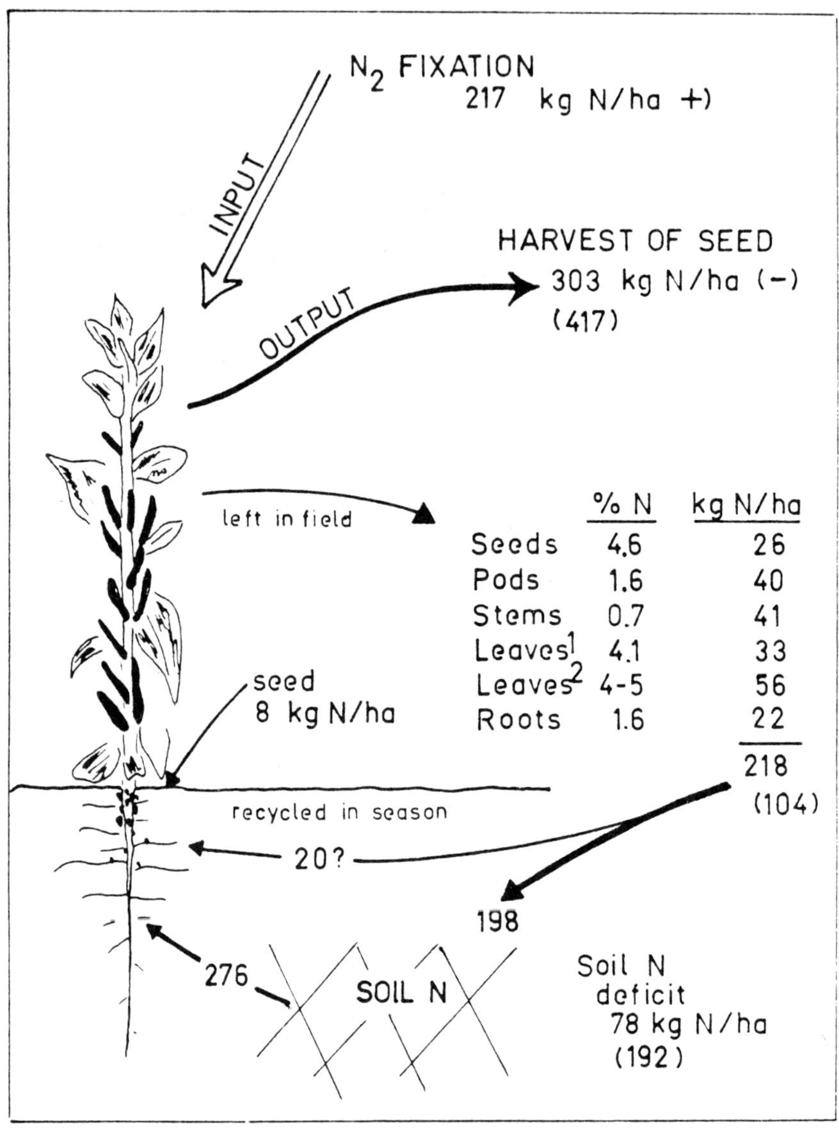

FIGURE 1   N budget for faba beans on Warren farm in 1978 (Patriquin *et al.*, 1981; Patriquin & Burton, 1982). The faba bean fixed 217 kg N from the atmosphere but 303 kg were removed in grains at harvest, indicating a net withdrawal from the soil of 78 kg (192 kg if straw were also harvested). The large amount of N taken up from the soil (276 kg) may have come in large part from recovery of N leached to deeper horizons prior to 1978 when normally fertilized cereal and maize crops were grown at this site.

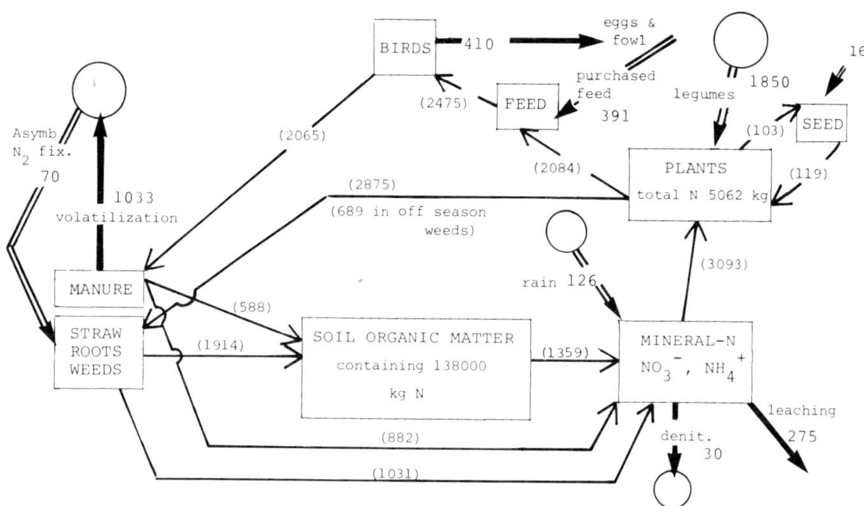

FIGURE 2   Nitrogen budget for the Aldhouse farm in 1979 (from Patriquin *et al.*, 1981). Numbers in brackets are flows of N between compartments within the farm. Big arrows and accompanying numbers not in brackets are flows of N into and out of the farm. Circles represent the atmosphere. Units are kg N per farm per year.

(2) For the farm as a whole, the inputs of N exceeded the outputs, even when the inputs in purchased grain are discounted. This illustrates that the deficiency of N for cereals was related to the manner in which N cycled around the farm, rather than to inadequate inputs.

(3) On the average, there was more N going into the fields than was being removed. Inputs to or outputs from the fields due to rain, asymbiotic N$_2$ fixation, leaching, denitrification and seeding were relatively small and roughly added up to zero; thus the major determinants of field N balances were the amounts of manure-N applied to fields, of N$_2$ fixation, and of crop-N removed at harvest. For bean fields, N$_2$ fixation (165 Kg) was approximately equal to the N removed at harvest (162 Kg). For the cereal fields, the inputs in manure (105 Kg) greatly exceeded the average outputs in grain (40 Kg). In other words the "N problem" appeared to be one of unavailability of most of the manure-N.

This excess input of N does not just disappear. If inputs exceed outputs, then the store of soil (humic) N, and the amount of N mineralized from this store each year, should increase each year until the outputs equal the inputs (Magdoff, 1978). The pertinent question is: how long would it take for this store of N to build up to a level at which good yields of cereals could be obtained? Some rough calculations indicated that we could expect cereal yields to increase by 11%, and bean yields by 2% in twenty years. Even allowing for conservatism in making these estimates, a "wait and let soil fertility accumulate" strategy was not a practical proposition (Patriquin *et al.,* 1981).

RESTRUCTURING THE SYSTEM

At this point, we considered that we needed to restructure the system in such a way that more of the N would cycle through crops, and less through weeds and humus; to make the system do more work for us at its present level of accumulated fertility; to take care of it in such a way that the release of N from humus, residues and manure coincides more closely with crop growth; overall, to increase the ecological efficiency of the system. Five areas that we have looked or are looking at in this regard are discussed below.

Crop rotation

In 1980, Mr. Aldhouse instituted a regular rotation of crops: Faba beans—oats underseeded with clover–clover–winter wheat. The clover is rotovated in the third year prior to planting winter wheat. Clover and winter wheat provide winter cover on the fields after the oat and clover crops respectively. Straw and weeds provide cover in the other two years. It is a cereal-legume rotation; legumes follow cereals so that the immobilizing properties of straw, and possibly carbon dioxide release (Shivashankar & Vlassak, 1978) stimulate N_2 fixation, and cereals in turn mop up N from decomposing legume residues. In a monoculture of cereals straw is frequently burned because of its immobilizing properties (Lynch, 1984). In this system those properties are a benefit. The faba bean is especially suited to this sequence because it begins to fix N shortly after germination when the wheat straw is likely to be immobilizing N, and it can use soil N during pod-fill when immobilization has likely ceased (Patriquin *et al.*, 1981; Patriquin & Burton, 1982).

In order to have as close as possible to one quarter of his farm in each stage of the rotation, Mr. Aldhouse brought two more fields into production, increasing the field crop area from 30 to 34.5 ha. For this system the calculated inputs to the fields from manure (1034 Kg) and N fixation (1858 Kg) exceed the outputs in grains (2087 Kg) by 805 Kg. Cycled into cereal grains at 2% N, this excess represents a potential increase in cereal yields of approximately 2 tonnes/ha, which is well above what is required for the yields to be similar to those achieved under conventional management.

Mineral nutrients and pH

Except for a few tonnes of lime applied in 1976, no fertilizer or lime has been used on the farm since 1976, and heavy liming has not been practiced since the sixties. Analyses of all fields in 1980 and most in 1983 indicated generally satisfactory base saturation (average 73%), no deficiencies in P, Ca or Mg, and slightly low K on 4 of 14 fields. Ratios of Ca:Mg are low (average 2.6) compared to those frequently cited as desirable (Albrecht, 1975).

Comparison of these and other data obtained since 1971 suggests that pH increased and stabilized at desirable values after Mr. Aldhouse stopped using fertilizer and lime, and that Ca and Mg increased in surface horizons by factors of about 20 and 50% respectively (Fig. 3). These changes are remarkable given that this is a high rainfall region of naturally acid soils. We believe that the changes are associated with, firstly, more complete cycling of N within the farm, and with the major inputs and outputs now being in non-ionic form (Helyar, 1976); and, secondly, enhanced vertical cycling associated with greater abundance of deep rooted herbs, particularly *Taraxacum officinale,* and possibly with more faunal activity. The herb and faunal effects could be related. Pfeiffer (1974) noted a close association of earthworms with *Taraxacum,* and we have noted the same phenomenon.

Cultivar selection

Comparisons of 6 oat cultivars illustrated that cultivars used in systems of biological husbandry, need to be selected in the same systems (see Fig. 5 and accompanying discussion below). On the basis of those comparisons, Mr. Aldhouse began to use the Fundy oat cultivar in 1984. We have yet to make comparisons of different wheat and bean cultivars.

Manure

There is a finite amount of manure coming out of the barn, and certainly less than we would need to relieve all N shortages immediately. Thus we would like to apply more manure where it imparts greatest benefits, and less elsewhere. To do so, we require estimates of the sustainable output of manure, and information on the variation in response to manure according to the crop (wheat or oats) and the particular field.

Information from the N balance (Patriquin *et al.,* 1981), a P balance, and from the literature (Patriquin *et al.,* in preparation) indicates that the sustainable output of manure is approximately 55 tonnes/year containing 1034 Kg N after volatilization losses. We are planning some trials on the use of

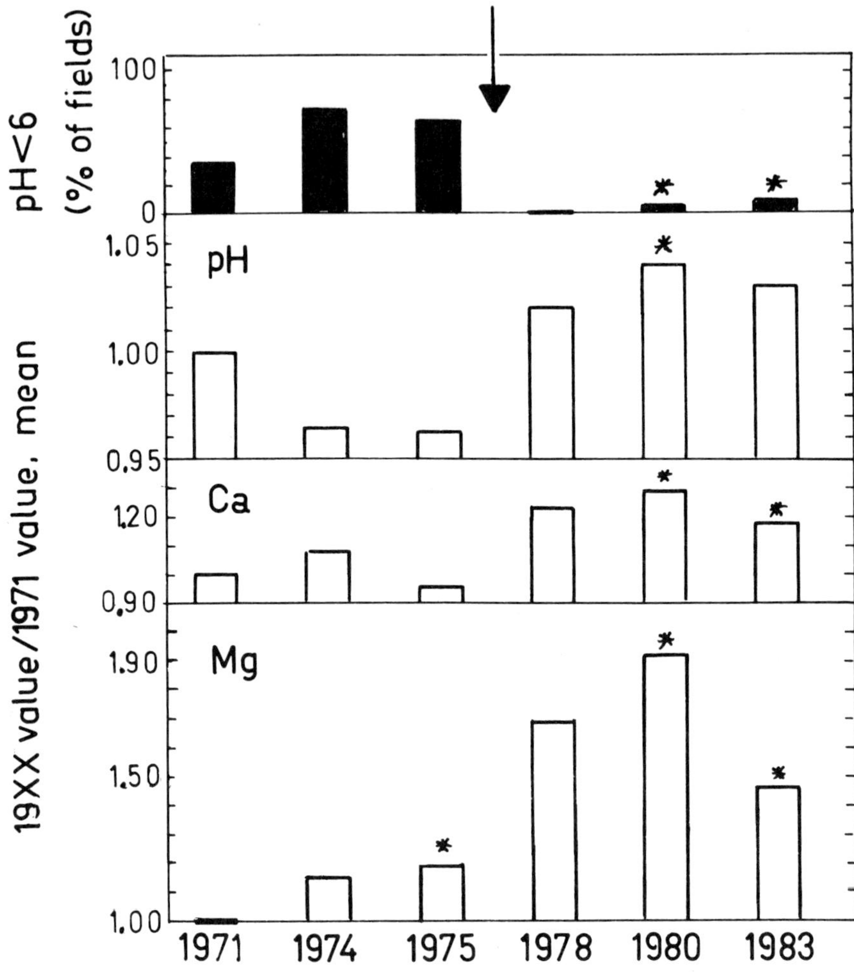

FIGURE 3   Summary of comparable data from analyses of soils taken from the Aldhouse farm, 1971–1983 (Patriquin *et al.*, in preparation). Arrow indicates cessation of fertilizer, lime and herbicide applications. Asterisks indicate significant differences ($\alpha = .05$) from 1971 values as assessed by binomial theorem for proportions, or rank- sum test for ratios. All (14) fields were sampled in 1971 and 1980. Seven fields were sampled in 1974, 11 (pH) or 12 (Ca and Mg) in 1975, 5 in 1977 and 10 in 1983. Samples taken from 1971 to 1978 were analyzed by the Nova Scotia Department of Agriculture, 1980 samples by the Woods End Laboratory, Temple, Maine, and the 1983 samples in our laboratory. Average values in 1971 were 6.06 for pH, and 6.41 and 2.08 meq/100g for Ca and Mg respectively.

gypsum to reduce volatilization losses in the roost (Roberts, 1897). Reduction of these losses by 50% would save about 500 Kg N (Fig. 2). Gypsum will be used because we also wish to increase the Ca/Mg ratio without increasing pH.

When Mr. Aldhouse stopped using fertilizers, he had an excess of manure available and applied it to all cereal fields at a rate of 5.6 tonnes/ha. As our calculation suggested this rate could not be sustained, after 1979 he stopped applying manure to oats which is considered the less demanding of the two cereals. From 1980 to 1983, wheat yields were about 30% below those achieved under conventional management. This degree of reduction in wheat yield appears to be typical of organic systems (Berardi, 1978; Lockeretz *et al.*, 1981).

Lockeretz *et al.* (1981) found yields of oats under organic management to be similar to those under conventional management. On the Aldhouse farm, however, oats yielded about 50% of normal from 1980 to 1983 with the exception of a good yield on one of the more fertile fields in 1981. In 1982 we applied 17 different combinations of fertilizer elements to plots on the oat fields. There was a consistent and large response to manure, and erratic responses to treatments which included N fertilizer. From these and other observations (discussed below) we concluded that the poor yields of oats are related in large part to phytotoxic effects of the bean residues. Manure may relieve this problem by feeding microbes as well as the plant, the microbes in turn breaking down the phytotoxins. Thus application of manure to oats appears to be one means of solving the problem. We do not know precisely how much is required, and how this requirement varies field by field.

In an attempt to begin to sort out the factors involved in field to field variation in response of crops to manure, we examined the mineralization of N and C in experimental soil-sand systems with or without manure or straw (Baines, 1984). There was substantial variation between soils from different fields in the amount of additional carbon dioxide or nitrate released or immobilized when residues were added (Table 1). The three soils with highest respiration (A4, A5, B2) in the presence of manure mineralized more additional N in the presence of manure than did three soils (A1, A2, C) with lower respiration values. The same three soils also exhibited the highest respiration when straw was added, and with the exception of soil A4, those three soils exhibited less immobilization of N than did the soils with lower respiration values. Assuming that the N required for microbial growth is proportional to the carbon dioxide output (Paul & Juma, 1981), one would expect the reverse, i.e. soils with lower respiration values to immobilize less in the presence of straw, and (possibly) to mineralize more in the presence of manure. A possible explanation is suggested by the straw carbon dioxide values, which for the three soils of highest activity, exceed the amount of carbon added as straw by substantial amounts. This suggests that there was substantial priming of the soil humus in soils with higher respiratory activity,

TABLE 1

Mineralization of C and N by soil from different fields without and with added residues (from Baines, 1984). Asterisks indicate soils for which the amount additional CO_2-C evolved in the presence of straw exceeded the amount of carbon added as straw (2250 μg C/g soil). Data are from Baines (1984).

FIELD	CARBON DIOXIDE-C			NITRATE-N		
	Soil alone	Additional CO_2 in presence of		Soil alone	Additional NO_3-N in presence of	
		straw	manure		straw	manure
		(μg CO_2-C or NO_3-N/g soil in 98 days)				
B2	1422	3114 **	2624	59	−26	112
A5	1985	3053 **	2650	67	−24	116
A4	2806	2812 **	2880	143	−53	95
C	1408	2122	1884	101	−32	54
A2	1337	1946	1566	91	−35	68
A1	1324	1867	2282	109	−58	84

Each sample consisted of 150 g air dried, sieved (1 mm mesh) soil mixed with 450 g quartz sand and 56 ml water in a 1.5 liter jar. Jars were closed with polyethylene and incubated at 30°C. Water was added as necessary to maintain the initial level. After 2 weeks (time zero), residues (0.75 g oat or 1.0 g manure) were added. Each treatment was replicated 3 times. For measurement of CO_2 production, jars were aerated for 30 minutes, closed, and and CO_2 measured after 19 hours. Cumulative values of CO_2 production were calculated from rates measured at 1, 2, 6, 7, 14, 28, 42, 70 and 98 days. For measurement of nitrate, 10 g soil + sand were removed and analyzed at 0, 2, 6, 14, 28, 59, 98 days.

N added in straw was 36 μg/g soil; N added as manure, 252 μg/g soil.

resulting in release of N and consequently in lower apparent immobilization and higher apparent mineralization of manure-N than in soils with lower respiratory activity. Regardless of the precise mechanisms involved, these observations suggest that the more biologically active the soil, the lower will be the amount of manure required to augment the N supply by a given amount, and the less immobilization there will be when low N residues are incorporated in the soil.

In the absence of a fully formulated analytical understanding of the variation in response to manure by crop and field, Mr. Aldhouse is continuing to apply manure to wheat at standard rates, and to oats as the excess allows. In time, this will give us an empirical assessment of the benefits of manure on each field.

Tillage

In terms of the biology involved, tillage is probably the most complex, most important and yet least understood of farming operations. On the Aldhouse farm, tillage operations are conducted with at least the following objectives in mind:

1. to facilitate good surface drainage and accordingly, rapid warming of the soil in spring;
2. to eliminate standing weeds, and to reduce seed banks but not to the point that weeds cannot function as a self-seeding cover crop;
3. to break up and incorporate the large amounts of residues from the wheat and bean crops in order to dissipate them and to encourage biological activity so as to minimize possible phytotoxic and immobilizing properties of these residues;
4. to incorporate the green manure (clover) crop prior to planting winter wheat;
5. to break up hard-pan smears left by shallow tillage;
6. to distribute straw in such a way that uniformly good drainage characteristics develop;
7. to break up surface soil and leave some residues near the surface so as to encourage capillary rise of water;
8. to encourage biological activity and release of mineral-N at the most appropriate time. In effect we are trying to sheet compost the residues, and want to provide as near as possible optimal conditions of air, water and temperature for the decomposers.

At the same time, we wish to minimize the well known negative effects of tillage including:

(i)   leaving the surface bare and subject to erosion;
(ii)   compaction;
(iii) use of fuel, time and labor.

It can be appreciated, I think, that even given all of the analytical information we asked for, it would be exceedingly difficult, if not impossible, to conceptualize precisely how the multitude of factors involved in tillage interact. The problem is additionally complicated by the presence of at least four distinct soil series. Yet decisions about the timing, frequency and type of tillage had to be made. In such circumstances, the farmer has to make as reasoned a guess as possible, try it out for a number of years, observe the effects and then adjust or try new techniques. In this regard it is the farmer who is the experimenter, and the scientist's role is primarily that of an observer and interpreter (L.H. Bailey in Roberts, 1897). Following is a brief account of how tillage operations have developed over the past five years.

Prior to 1976, Mr. Aldhouse mouldboard ploughed his fields regularly in

the fall. He stopped doing so after taking up biological husbandry because of
(i) the difficulty of ploughing after high residue crops (such as wheat and
clover), (ii) the difficulty of contour ploughing on irregularly shaped fields;
and (iii) the probable ill-effects of burying residues in a layer at depth
(Faulkner, 1945). Since 1976, tillage has consisted primarily of rotovating
residues into the ground in the fall or spring after beans, in September after
wheat, and in June or July after clover. The seedbed is then prepared by
harrowing with a spring tooth harrow, and according to the weediness of the
field, it may be gone over once or twice with a spike-tooth harrow after the
crop is planted.

In 1979, Mr. Aldhouse decided not to conduct tillage operations in the fall
because of the possibility that to do so would encourage leaching and erosion.
In the spring of 1980, there was a very heavy growth of weeds following the
previous year's bean crop. The field was rotovated about two weeks prior to
planting oats. The oat yield that year was exceptionally poor, which we
attributed to phytotoxic effects of the decomposing weed residues. In order to
avoid this problem, Mr. Aldhouse then decided to rotovate the bean residues
in the fall which is more desirable with regard to the seasonal distribution of
labor and to the workability of the land. We think that this practice does not
cause excessive leaching or erosion because (i) the bean crop is harvested in
October by which time soil temperatures have begun to drop (i.e. there should
not be a lot of decomposition before winter sets in); (ii) laboratory studies
suggested that the bean residues would immobilize N initially (Patriquin *et al.*,
in preparation), (iii) the surface is left rough, and there is still a fair amount of
weed growth.

In spite of these adjustments, oat yields remained poor over the next three
years, with one exception, that being on one of the most fertile fields in 1981.
Data from an oat cultivar experiment (Fig. 4) suggested that even with fall
tillage, we had problems with phytotoxicity—presumably from the bean
residues, since there was not now a heavy growth of weeds. A solution to this
problem could lie in applying manure to the oats, as discussed above, but
alternative tillage operations may provide the most appropriate solution.

Mr. Aldhouse considered that rotovation of residues is not completely
satisfactory because while it mixes residues into the soil, it leaves a fair amount
on the surface, and leaves the surface flat which tends to keep it cold. In the fall
of 1983, he tilled one field after beans with a tool bar equipped with six right
hand throw shanks and 3-inch shovels. This left the soil surface in nicely
ridged condition (spaced at 14 inches) and effected good mixing of residues
within the soil without actually turning the soil over—or with residues
occasionally lumping together as they are liable to do with chisel ploughs. The
oat crop that developed on this field in 1984 was a good one (oat biomass, 5144
kg/ha; compare with Fig. 4) and the yield on a manure-fertilized section was
similar to those on the non-fertilized section (Patriquin *et al.*, in preparation).

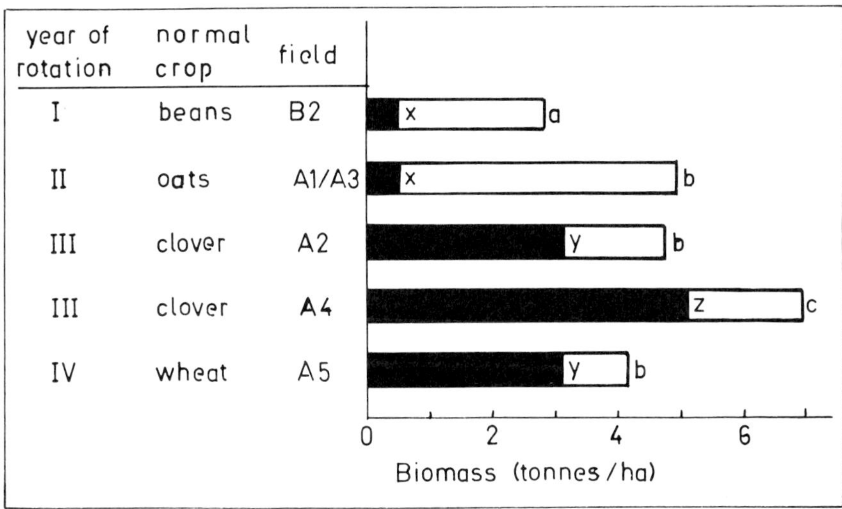

FIGURE 4   Total biomass (crop + weeds) and crop (shaded) biomass on oat cultivar test plots in 5 fields in 1982. Bars followed by different letters represent yields which differed significantly from each other as assessed by the rank sum test ($\alpha = 0.05$). On each field, six cultivars were planted, each cultivar in three 2×2m plots. The central 35×35cm were harvested at maturity. The low oat yields on field B2 and A1/A3 are indicative of phytotoxic effects of residues from the previous years' crops (wheat and beans respectively). The low total yield on field B2 is attributed to immobilization of N by wheat residues. Data from Patriquin *et al.* (in preparation).

Likewise, a good oat crop (oat biomass 4718 kg/ha) grew on another field which Mr. Aldhouse mouldboard ploughed in the fall of 1983. In both cases, we think that the apparent improvement over the previous situations is related to improved surface drainage and more rapid warming of the soil in the spring, that resulting in turn in accelerated biological activity and breakdown of phytotoxins.

At this point, tillage operations on the farm are planned as follows: (i) to rotovate in the fall after beans, followed by ridging with twisted shovels; (ii) to rotovate after wheat, in the fall (there is too much residue for the ridging operation); (iii) to "partial summer fallow" in year III after clover by deep rotovating (with rear doors left up so that weeds thrown on the surface remain there), and use of twisted shovels and harrows. These operations will be followed by harrowing for seedbed preparation and weeding. We consider that a partial summer fallow is necessary to control perennial weeds, particularly Canada thistle (*Cirsium arvense*), which have increased in abundance over the past four years (unpublished data). Cultivations will begin in early June and will be spaced at approximately 21 day intervals, which has been found effective for control of Canada thistle (Hodgson, 1958), until wheat is planted in late August. Clover reaches near maximum biomass by early June. As we are following it with a long season crop (winter wheat), we are hopeful that most of the N mineralized during the fallow will be recovered.

ANNUAL WEEDS AND THE CONSERVATION OF N

In a recycling system, the conservation of even small amounts of N is of great importance. Present losses due to denitrification and leaching are of the order of 10 kg N/ha and are approximately balanced by inputs in rain, asymbiotic N fixation and seed. If the losses increased, then the sustainable output of grain N would decline accordingly. In terms of the soil N balance, 20 Kg N lost by leaching is equal to a grain output of about 1 tonne/ha. Off-season growth of weeds on the farm conserve of this order of magnitude of N, and thus are of critical importance.

Note by contrast that such amounts of N have much less significance in conventional, more open systems. For example, given a grain crop with 60 Kg N in the grain, 30 Kg N in the straw which is harvested or burned, and 50% loss of fertilizer-N, the N requirement is 180 Kg N, and a saving of 20 Kg N represents a saving of only 0.3 tonnes grain—roughly we can say that N is three times more valuable in the recycling system.

In addition to their role in conserving N, weeds protect the surface of the soil, and fix carbon and some, N_2, where or when crops are not present, and bring up nutrients from deep horizons. They may play a critical role in insect pest control (Altieri & Whitcomb, 1978/79). Thus the strategy sought with regard to weeds is to control them so that they do not interfere with crops, but not to eliminate them so that they are always available as "self-seeding cover crops". This strategy, and the discussions following, apply to annual (and biennial) rather than to perennial weeds, because the annuals are easier to control, compete less with the crop, and by their presence, help to control the more problematical perennials (i.e. perennials would be much more of a problem in the total absence of annuals).

The key to this strategy lies in the epigenetic relationship (Thomas, 1983) between weeds and crops: crops have a negative effect on weeds and weeds a negative effect on crops. In such a relationship, whoever gets a head start will hold the upper hand. Thus we can control weeds both by giving a helping hand to the crop or by hindering the weeds.

Of the many factors influencing the crop-weed relationship, the most important are crop rotation and cultivation, which hinder the weeds, and fertility, which has a positive effect on the crop.

Each crop has a characteristic assemblage of associated weeds, and changing the crop each year helps to keep prevent any one species from building up to the point that it cannot be controlled. For example, cultivation of soil for winter wheat in the fall stimulates wild radish (*Raphanus raphanistrum*) the most abundant summer annual weed, to germinate. The radish grows between rows of wheat in the fall, protecting the soil and conserving nutrients, but is killed over winter, resulting in a substantial reduction in the seed bank (Patriquin *et al.*, 1981).

The critical question with regard to cultivation of annual weeds is: how much is enough, and how much is too much? A completely clean soil is undesirable as is a crop overgrown by weeds. Our studies on fertility-weed interactions have provided us with a tool for looking at this question.

We argued that given an epigenetic relationship and provided the crop gains the initial advantage, the higher the fertility, the better the crop will do and the fewer weeds there will be at harvest. An analysis of quadrat data from three crops and several farms supported this concept (Patriquin *et al.*, 1981). We assumed that the total biomass is a relative measure of fertility (i.e. something will grow whether it is weeds or crop). Thus we would expect that the higher the total biomass, the fewer weeds there should be in the sample. Four types of relationships were observed (Fig. 5). The type 4 relationship appears to represent the limit of permissable weediness for beans, i.e. it represents a situation in which the crop holds the upper hand and the weeds fill in all available spaces between the crop but do not overwhelm the crop. Comparison of bean yields on weeded and unweeded plots support this concept; yields on weeded plots averaged only 9.7% higher than those on unweeded plots (and the differences were not statistically significant). There was no trend of increasing advantage for the weeds at low total biomass when the crop to weed ratio is low. Oat yields on untreated plots in a field where the relationship between percent crop and total biomass was of type 3 were 17% below those of herbicide treated plots (Patriquin *et al.*, in preparation). Relationships of type 5 clearly represent situations in which the weeds have gained the upper hand. The relationships are obvious visually: where the crop has the advantage, annual weeds predominate only in regions of poor (short or sparse) crop growth.

The relationships have fundamental practical significance for biological husbandry: (i) they illustrate that a high proportion of annual weeds is "normal" at low fertility, and even desirable; (ii) they indicate that problems with annual weeds should decline as fertility increases, and finally (iii), relationships of type 5 illustrate situations in which more cultivation is required. For example, for wheat fields over the last four years, we have noted a tendency for a shift from a type 3 to a type 5 relationship, suggesting that more control of weeds is required.

We have used the same approach to analyze the performance of different cultivars of oats. Such analyses illustrated that certain modern varieties, selected under conditions of few or no weeds, are not competitive with weeds

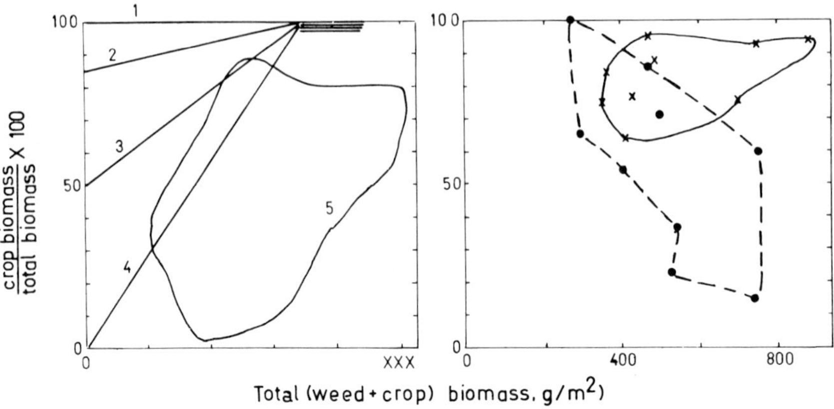

FIGURE 5   Relationships between the proportion of crop in a sample taken at harvest (expressed as percentage) and the total (weed + crop) biomass (Patriquin *et al.*, 1981 and in preparation). Relationship (1) is characteristic of fields in which there is nearly total elimination of weeds by herbicides. Relationships 2, 3, and 4 represent fields in which there was a significant positive linear correlation between percent crop in a sample and total biomass. Relationship (2) was observed for oats in 1979 following a cultivated fallow and for wheat in 1980, relationship (3) for wheat (most years) and for oats in 1981 and 1984, and relationship (4) for faba beans (most years). Relationship (5) represents faba beans on N fertilized plots, oats in 1980, 1982 and 1983, and one wheat field in 1984.
FIGURE 5 Right illustrates these relationships for the best performing oat cultivar (a traditional type) and the most poorly performing oat cultivar (a modern selection) in oat cultivar comparisons, excluding fields following faba beans or wheat (see Fig. 4 above).

(Fig. 5). Unfortunately, some of the older varieties are now difficult to obtain. In any case, if biological husbandry is to progress, cultivars with the benefits of other modern cultivars such as high harvest index need to be selected *de novo*.

IS NITROGEN LIMITING?

It is generally assumed that if a crop responds to fertilization with a certain mineral, then the mineral concerned is "limiting" for plant growth. This may be true for the plant considered alone, but it does not necessarily follow that adding more of the element is the most appropriate way to solve the limitation; or, for example, that the quantity of N in the system is insufficient to support higher yields. The limiting factor concept becomes especially clouded when we are dealing with cyclical processes, because we must then ask what is "limiting" production of the mineral by the previous step in the cycle, and then the previous step to that . . . and so on.

Even in a narrower context, the limiting factor concept can be misleading. In 1980, we observed a pronounced response of wheat on field A1 to N fertilizer applied in 2 × 2 m plots. Since the standing amount of inorganic N in the field at large was small compared with that taken up by wheat, we supposed that variations in growth and N accumulation by wheat in that field would be related to the N mineralization capacity of the soil. A comparison of plant N accumulation with the soil N mineralization potential suggested that the former was indeed related to the latter, but that other factors were also limiting (Fig. 6). What are those other factors? At least one of them involves drainage.

In 1983, we measured plant height and soil matric potential during a saturating rainfall at 25 randomly chosen sites, and at 12 sites of adjacent tall and short wheat in each of 3 fields. For the randomly chosen sites there was no correlation between height and soil matric potential. However, at 12/12, 11/12 and 9/12 of the paired sites, soil matric potential was lower in the stand of taller growth (Patriquin *et al.*, in preparation), i.e. the stands of taller growth drained more rapidly. This suggests that over the field(s) at large, variation in growth is related to variation in potentially mineralizable-N, but within regions of the field, to drainage and possibly other factors.

What is "limiting" drainage? The stands of tall and short wheat tended to be oriented parallel to each other and in the direction of operation of the combine and to be separated by approximately the width of the combine. This suggested that variations in drainage are related to the pattern of straw distribution and/or compaction caused by passage of the combine. Examination of several sites revealed obvious straw residues in soil blocks from stands of good growth but not in those of poor growth, confirming that the regions of poor growth at least include some of those where little straw is laid down and where compaction may be greatest.

In 1980, yields on all six N-fertilized plots exceeded those of controls, and averaged 2.6-fold higher. How then could drainage also have been a limiting factor? Would simultaneous improvement of drainage have increased yields in the presence of N fertilizer even further? The answer is probably no. The effects of drainage are not independent of those of N. Drainage affects the efficiency of N use by the wheat (Armstrong, 1980), the efficiency being higher in better drained soil. It might also affect the actual mineralization (i.e. the degree to which the potential mineralization is actually realized); the literature is not very clear on this point. When wheat is fertilized at levels found to be necessary for uniformly high yields (the usual criterion), the effects of variation in drainage are essentially compensated for—and unless one is actually measuring the efficiency of fertilizer use, or fertilizing at submaximal levels, go unnoticed. More poorly drained wheat uses N less efficiently, but since excess N at sites in which it is being used efficiently by the plant is probably lost by leaching or denitrification anyway, these differences in

                                    D.G. PATRIQUIN

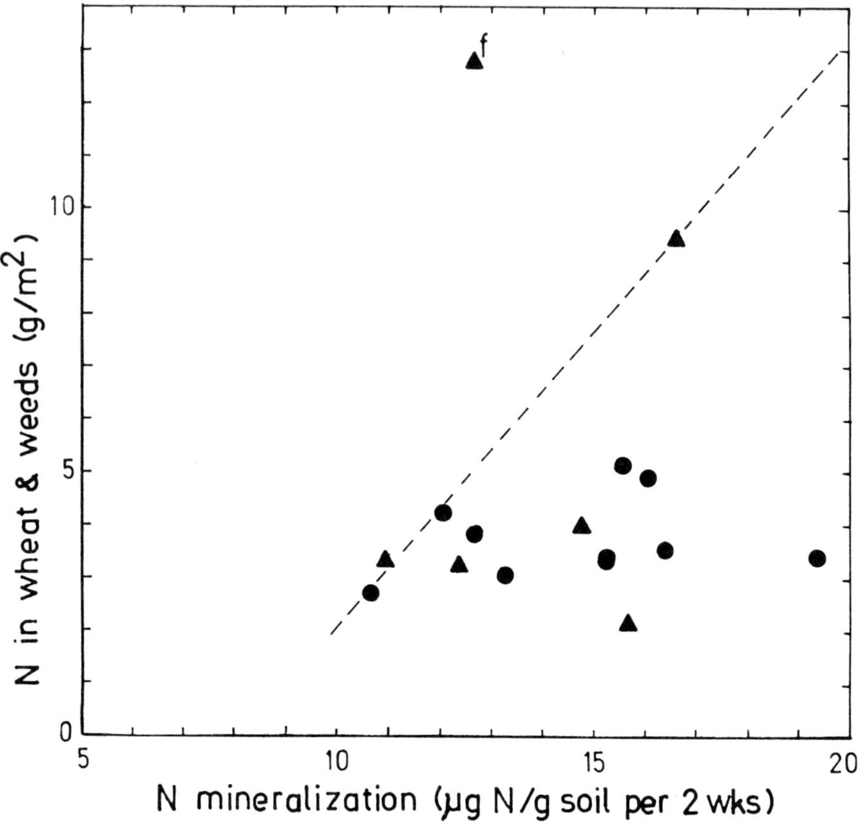

FIGURE 6   Relationship of total N in winter wheat and weeds on June 4, 1980, and mineralizable-N (from Patriquin *et al.*, in preparation). Circles represent samples from randomly chosen sites; triangles represent samples taken from sites chosen to include a wide range of productivity. Triangle "f" represents a plot fertilizer with 115 kg N/ha as urea on May 18. The broken line links sites at which N was limiting; points below this line represent sites at which the maximum potential for N accumulation was not realized because of the operation of other limiting factors (for discussions of this sort of interpretation, see Balandreau & Ducerf, 1980; Parnas, 1975). Mineralizable N was measured by a laboratory incubation technique on 150 g soil from each site.

efficiency are not normally evident. One of the most striking differences between fields managed by biological husbandry and those managed chemically is the much greater variation in growth in the former, at least during the transition to biological husbandry, i.e. when the masking effects of fertilizers are initially removed.

The interaction between drainage and N illustrates an important point: if we diagnose our systems purely in terms of chemistry, we will come up with only chemical "solutions". The problem with such solutions is that by masking or compensating for other limitations or inefficiencies in the system,

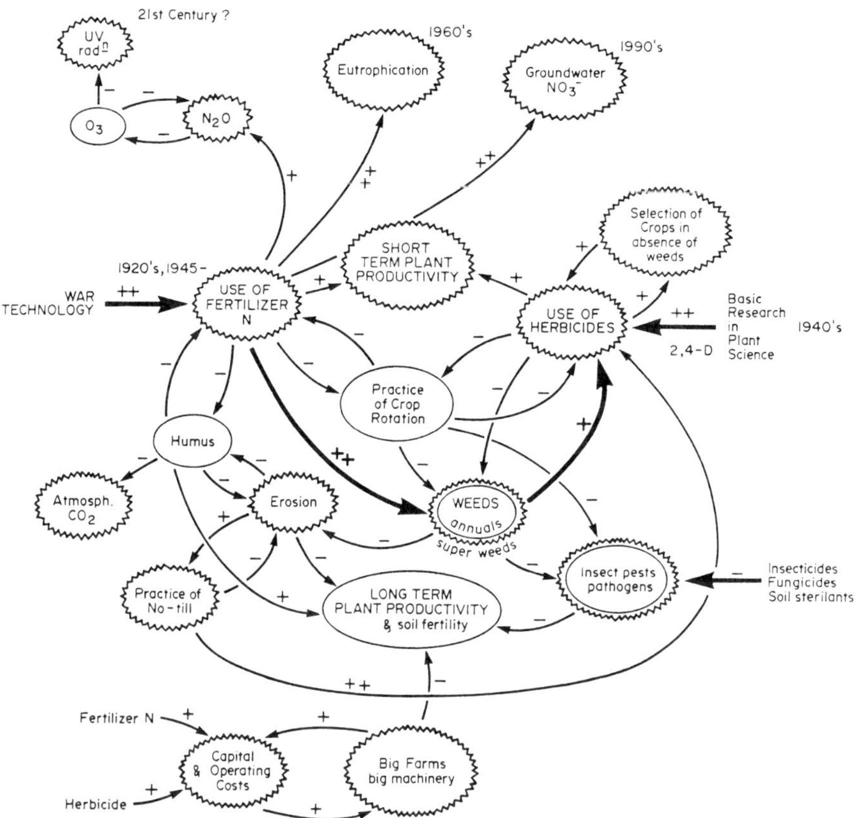

FIGURE 7   Some positive feedback loops generating dependence on N fertilizer and herbicide, and resulting in deterioration of the environment and increased cost of production. The net effect (positive and negative) of increase or intensification of one variable or process on another or on itself via a given pathway is given by the parity of the number of negative interactions on that pathway; if the number is odd, the net effect is negative, if it is even, the net effect is positive (Meadows *et al.*, 1972; Thomas, 1983). Hatched enclosures represent processes or variables which have intensified or increased, and smooth enclosures, those which have slackened or decreased since the 1940's. The author suggests that initiation of widespread use of fertilizer-N and herbicide in the forties was related to these materials suddenly becoming readily available rather than to real requirements for them. Their use in turn, generated real requirements via the illustrated pathways. For example, use of N fertilizer stimulates weeds; herbicides are used to control weeds eliminating the relatively innocuous annuals which normally protect the soil after harvest; erosion increases, there is loss of humus and the requirement for N fertilizer increases.

they lead to deterioration of these other factors, and to greater dependence on chemicals. The high degree of correlation between cereal yields and fertilizer use is repeatedly cited as evidence that nutrients are limiting, and that more inputs of fertilizer, particularly of N are necessary for enhanced food production (e.g. Greenwood, 1982). By neglecting the factors leading to "N limitation", i.e. by dealing with the proximate rather than ultimate causes of N limitation, such diagnoses generate greater dependence on fertilizer (Fig. 7).

We are frequently asked if we would consider composting manure in the system. The answer is that we think there is no question that to do so would benefit the soil and improve cycling and productivity, both through its effects on soil structure, and on its biological activity. However, there is not excess carbon available for composting. Is the system then carbon limited? Should we grow some high carbon crops purely for composting? If the concept of limiting factors is valid at all in biological husbandry, it must be applied to the decomposers as well as to the producers, and be extended to non-chemical parameters of the system.

CONCLUSION

If there is a "N problem" in biological husbandry it is this: given a bag containing 100 kg fertilizer-N, we know exactly what to do with it. It may be wasteful, but we can pretty well guarantee that there will be no N shortage for the crop. However, given 100 kg N in soil humus, or in manure or in plant residues, we have very little idea of how to use it. If it is manure –N, we generally consider that only the N mineralized in the first year is available. If it is N in straw, we may burn it. Associated with graminoid crops, legumes are more often considered weeds than they are donors of N to the crop.

Many attempts have been made to relate the N-fertilizing values of organic materials to some chemical fraction of those materials, e.g. %N, N released after autoclaving etc. (e.g. Whitehead, 1981). In relation to the efficiency of use of fertilizer-N, these chemical indices may give acceptable results. For biological husbandry, however, such approximations are simply too crude, because the amount of N transformed is as much a function of living catalytic activity as it is of initial chemical composition of the materials themselves.

When Mr. Aldhouse stopped using N fertilizer on his farm, his cereals suffered from a severe shortage of N. By analogy with the use of industrially fixed N to overcome N shortages in conventional agriculture, we supposed that this shortage could be overcome by increasing biological N fixation. We promptly found that the problem lay not in the quantities of N entering the system, but in the way it cycled around the system. That in turn was not really a N problem as much as it was a complex of problems related to weed, residue and manure management; in effect of basic ecology. We finally began to recognize what is probably the most important principle in biological husbandry: productivity is intensified not by augmenting inputs but by intensifying cycling (Koepf *et al.*, 1977). This is achieved by maximizing the biological activity of all components of the systems. In the parlance of self-organization theory (Jantsch, 1980) the farm is a "hypercycle"—a cyclical process in which some of the stages are autocatalytic. The intensity of cycling in such systems is dependent primarily on the input of solar energy and on the catalytic activity. The catalysts are made up of the totality of the biological materials on the farm—the humus, the microbes, the soil fauna ... the livestock and man himself; each is at once a product, a precursor and a catalyst, and the well being of each depends on and contributes to the well being of the other.

This is not to say that all N deficiencies can be overcome by intensifying cycling. There is a certain minimum amount of N that must be present in a system to "create" a N cycle (Bradshaw *et al.*, 1982)—in effect to build up the catalytic material—and to a point, the more N that there is in the system, the greater will be the catalytic activity. The sustainable inputs of N to the system in turn determine its sustainable output as product after discounting losses such as leaching. But the important point is that the problem begins rather than ends with the input-output balance, as opposed to conventional systems in which the main concern is; how much N is enough? And from that point on, diagnosis of the system's "limitations" is essentially an ecological problem that must be approached with "all sensory and intellectual channels open" (Hill, 1982).

SUMMARY

Since 1978, the author has been conducting research into the theory and practice of biological husbandry in collaboration with a farmer who stopped using pesticides and mineral fertilizers in 1976. Eggs are exported from the farm. About 60% of feed is grown on the farm in a legume-cereal rotation (faba beans-oats-clover-winter wheat), and plant and animal residues are recycled. Annual weeds function as a self-seeding cover crop, protecting the soil, conserving nutrients and fixing carbon where and when cultivated crops are not present.

Yields average about 25% lower than those on conventional farms, but the farm is more profitable because of lower input costs. A nitrogen budget suggests that inputs of nitrogen are sufficient to sustain cereal yields equivalent to those of conventional systems. However, much of the annual input of N to cereal fields, in manure, is not available in the short term. Various laboratory and field studies suggest that as fertility or the biological activity of soils increases, problems related to immobilization of N by straw, phytoxicity and annual weeds decline, and that less manure is required to augment the N supply by a given amount. While N might be identified as the "limiting factor" for cereal production, alleviation of N shortages is dependent on intensifying cycling, rather than on increasing N inputs. This intensification is achieved by augmenting natural rhythms on the farm through appropriate tillage techniques, and by ensuring an abundance and high activity of the catalysts of the N cycle, i.e. of all of the farm biota.

ACKNOWLEDGEMENTS

The scientific work referred to in this paper was supported in part by the Natural Sciences and Engineering Research Council of Canada. I trust it is apparent that this work owes most to the insight, labor and hospitality of Basil and Lillian Aldhouse.

References

Albrecht, W.A. (1975). *The Albrecht Papers* (C. Walters Jr., ed.). Acres U.S.A.; Raytown, Missouri.

Altieri, M.A. & Whitcomb, W.H. (1978/1979). Manipulation of insect populations through seasonal disturbance of weed communities. *Protection Ecology*, 1, 185–202.

Armstrong, A.C. (1980). The interaction of drainage and the response of winter wheat to nitrogen fertilizers: some preliminary results. *Journal of Agricultural Science (Cambridge)*, 95, 229–231.

Baines, D. (1984). Effect of different soil types on mineralization of nitrogen from agricultural residues. Honors Biology thesis, Dalhousie University; Halifax, Canada.

Balandreau, J. & Ducerf, P. (1980). Analysis of factors limiting nitrogenase (C_2H_2) activity in the field. In *Nitrogen Fixation*, Volume II (W.E. Newton & W.H. Orme-Johnson, eds), pp. 229–242. University Park Press; Baltimore.

Berardi, G.M. (1978). Organic and conventional wheat production; examination of energy and economics. *Agro-Ecosystems* 4, 367–376.

Bradshaw, A.D., Marrs, R.H., Roberts, R.D. & Skeffington, R.A. (1982). The creation of nitrogen cycles in derelict land. *Philosophical Transactions of the Royal Society of London*, B296, 557–561.

Candlish, E. & Clark, K.W. (1975). Preliminary assessment of small faba beans grown in Manitoba. *Canadian Journal of Plant Sciences*, 55, 89–93.

Earl, C.D. & Ausubel, F.M. (1983). The genetic engineering of nitrogen fixation. *Nutrition Reviews*, 41, 1–6.

Evans, L.E., Seitzer, J.F. & Bushuk, W. (1972). Horsebeans—a protein crop for Western Canada? *Canadian Journal of Plant Sciences*, 52, 657–659.

Faulkner, E. (1945). *Ploughman's Folly*. Michael Joseph Ltd., London.

Greenwood, D.J. 1982. Nitrogen supply and crop yield: the global scene. *Plant & Soil*, 67, 45–59.

Harmsen, G.W. & van Schreven, D.A. 1955. Mineralization of organic nitrogen in soil. *Advances in Agronomy*, 7, 299–337.

Helyar, K.R. (1976). Nitrogen cycling and soil acidification. *Journal of the Australian Institute of Agricultural Science*, 42, 217–222.

Hill, S.B. (1982). Steps to a holistic ecological food system. In *Basic Technics in Ecological Farming*, (S.B. Hill and Pierre Ott, eds.), pp. 15–21. Birkhauser Verlag; Basel, Boston, Stuttgart.

Hodgson, J.M. (1958). Canada thistle (*Cirsium arvense* Scop.) control with cultivation, cropping, and chemical sprays. *Weeds*, 6, 1–11.

Jantsch, E. (1980). *The self-organizing universe*. Pergamon; Oxford, New York.

Johnson, J.W., Welch, L.F. & Kurtz, L.T. (1975). Environmental implications of N fixation by soybeans. *Journal of Environmental Quality*, **4**, 303–306.

Koepf, H., Pettersson, B.D. & Schaumann, W. (1976). *Biodynamic Agriculture. An Introduction* The Anthroposophic Press; New York.

Lockeretz, W., Shearer, G. & Kohl, D.H. (1981). Organic farming in the corn belt. *Science*, **211**, 540–547.

Lynch, J.M. (1984). Interactions between biological processes, cultivation and soil structure. *Plant & Soil*, **76**, 307–318.

Magdoff, F.R. (1978). Influence of manure application rates and continuous corn on soil-N. *Agronomy Journal*, **70**, 629–632.

Meadows, D.H., Randers, D.L. & Behrens, W.W. (1972). *The Limits to Growth: A Report for the Club of Rome's Project on the Predicament of Mankind.* Universe; New York.

Parnas, H. (1975). Model for decomposition of organic material by microorganisms. *Soil Biology & Biochemistry*, **7**, 161–169.

Paul, E.A. & Juma, N.G. (1981). Mineralization and immobilization of soil nitrogen by microorganisms. *Ecological Bulletins (Stockholm)*, **33**, 179–195.

Patriquin, D.G. & Burton, D. (1982). Faba bean: an alternative to soybean in Nova Scotia, Canada. In *Basic Technics in Ecological Farming* (S.B. Hill and P. Ott, eds), pp. 98–107. Birkauser Verlag; Basel, Boston, Stuttgart.

Patriquin, D.G., Burton, D. & Hill, N. (1981). Strategies for achieving self sufficiency in nitrogen on a mixed farm in eastern Canada. In *Genetic Engineering for Nitrogen Fixation and Conservation of Fixed Nitrogen* (J.M. Lyon, R.C. Valentyne, D.A. Phillips, D.W. Rains & R.C. Huffaker, eds.), pp. 651–671. Plenum; New York.

Patriquin, D.G., Hill, N., Baines, D., Bishop, M. & Allan, G. (In prepn.) Observations on a mixed farm in the transition to biological husbandry.

Pfeiffer, E.E. (1974). *Weeds and What They Tell.* Biodynamic Farming and Gardening Association; Springfield, Illinois.

Rice, W.A. (1980). Seasonal patterns of nitrogen fixation and dry matter production by clovers grown in the Peace River region. *Canadian Journal of Plant Scences*, **60**, 847–858.

Roberts, I.P. (1897). *The Fertility of the Land.* MacMillan; New York.

Shivashankar, K. & Vlassak, K. (1978). Influence of straw and CO on N-fixation and yield of field-grown soybeans. *Plant & Soil*, **49**, 259–267.

Stewart, W.D.P. & Rosswall, T., eds. (1982). The nitrogen cycle. *Philosophical Transactions of the Royal Society of London*, **B296**, 299–576.

Thomas, R. (1983). Logical description, analysis, and synthesis of biological and other networks comprising feedback loops. In *Aspects of Chemical Evolution* (G. Nicolis, ed.) p.247, Wiley; New York.

Thompson, R. & Taylor, H. (1982). Prospects for *Vicia faba* L. in Northern Europe. *Outlook on Agriculture*, **2**, 127–133.

Whitehead, D.C. (1981). An improved chemical extraction method for predicting the supply of available soil nitrogen. *Journal of Science of Food and Agriculture*, **32**, 359–365.

Mycorrhiza in a Sustainable Agriculture

Barbara Mosse

11, The Close, Harpenden, Herts, AL5 3NB, U.K.

INTRODUCTION

Before discussing the possible significance of mycorrhiza in a sustainable agriculture I would like to consider briefly some of the more important problems that seem to endanger the stability of present agricultural systems (Table 1). Some of these problems occur mainly in highly developed agricultural systems in temperate climates, others are characteristic of developing systems in the tropics and some apply to both.

TABLE 1

Problems endangering the stability of present agricultural sytems

Temperate:	Tropical:
Eutrophication	Erosion, laterisation, desertification
Acid rain	Salinity arising from irrigation
Biocides	Deforestation
Erosion, loss of organic matter, and degraded soils	

Eutrophication. This is due to rising nutrient levels in the water of rivers, lakes and reservoirs, to such an extent that marine life is endangered and the waterways become choked with vegetation. Such pollution can be caused by domestic and industrial effluents, and by run off and drainage water from highly fertilised fields.

Acid rain. This problem seems to be spreading in certain parts of Europe affecting mainly temperate forest trees. In Sweden it is also thought to endanger some annual crops. The problem is attributed to sulphur dioxide pollution of the atmosphere resulting in acid rain and a possible lowering of the soil pH. This can affect plant growth directly, but may also have even more serious indirect effects by increasing the availability of heavy metals.

Biocides. Pest and disease control in highly intensive systems of monoculture has entailed increased use of biocides, and more potent fungicides and pesticides are constantly being developed. As well as the immediately desired effects, such substances often cause changes in the entire ecosystem. Many nursery soils and larger areas used for intensive horticulture are regularly fumigated. Sometimes, as in the citrus nurseries of California, this is required by law. One third of the productive land of Israel is said to be fumigated regularly. Soil sterilisation to eliminate nematodes is practised regularly in Hawaian pineapple plantations. In addition to such *in situ* soil sterilisation many plantation crops, for instance coffee in Brazil, are raised in sterilised soils before outplanting. Even more extreme is the growing practice of aseptic clonal propagation used increasingly, for example for oil palms, asparagus and strawberries. This produces completely aseptic plants in the first instance, and the eventually outplanted material is colonised only by microorganisms derived from the air and not from the soil.

Erosion, loss of organic matter and degraded soils. Erosion can be caused by wind, more usually by violent rain storms or a combination of the two. It is basically associated with loss of soil structure caused by denuding the soil of vegetation, by overcropping, overstocking and loss of organic matter. Some soils are much more liable to such deterioration than others. For instance I have seen tea gardens in Sri Lanka on exceedingly steep slopes, subject to torrential rains and apparently suffering little from erosion. Equally the Rothamsted experience of continued good wheat yields and little soil deterioration after continuous application of mineral fertilizers for over 100 years owes much to the stable clay loam of Broadbalk field. Figure 1 illustrates the very slow mineralisation of organic matter in another Rothamsted soil. Many soils, however, do not possess such stability, and long term losses of soil organic matter in the prairie soils of N.W. America are causing considerable concern. In the tropics loss of organic matter is greatly accelerated by high temperatures. Once serious erosion occurs and the surface layer of the soil is lost, laterisation begins and the soil becomes cement-like and unable to support any plant growth. Thus desertification begins.

Degraded soils are the result of operations like open-cast mining, oil shale and gravel extraction and other major disturbances of the soil surface.

Salinity. As a result of irrigation and associated increases in surface evaporation high levels of Na and Cl can build up in the upper layers of the soil which may become toxic to some plants. Excessive water usage and consequent lowering of the water table may lead to seapage of sea water into the ground water supply. In Cyprus this has destroyed some citrus orchards.

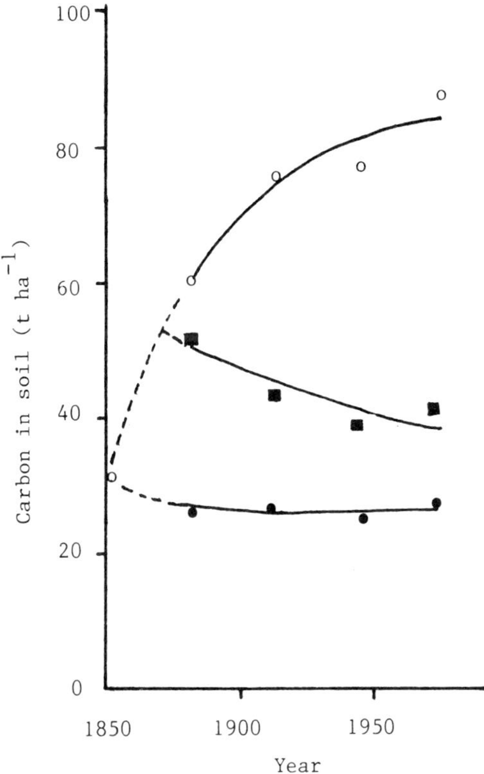

FIGURE 1 Organic carbon in the topsoil of Hoosfield continuous barley experiment at Rothamsted. o—received FYM annually; ■—received FYM annually between 1850–1871, thereafter; ●—unmanured. (From Jenkinson & Johnstone, 1976).

Deforestation and Afforestation. Forests are becoming an increasingly valuable resource. The progressive destruction, particularly of tropical rain forests, and subsequent problems of re-afforestation, or indeed any use of the denuded land, have caused much concern to many ecologically minded people. Often native tree species are being replaced by faster growing exotics, often species of *Pinus*. With these an understanding of mycorrhizal requirements can be crucial for establishment.

To illustrate these problems more clearly, I would like to quote some figures from two publications, 'Ein Planet wird geplündert', a book by Herbert Gruhl and an article entitled 'Maintenance and increased plant production in developing countries' by Prof. S. Rehm of Göttingen University. Between 1951 and 1966 world food production increased by 34%. This has been accompanied by the following increases, tractors 63%, PO_4 75%, NO_3 146%, pesticides 300% (Goldsmith *et al.*, 1972). Table 2 shows figures, calculated by Pimentel *et al.* (1973), of energy requirements for maize production in USA between 1945 and 1970. 60–80% of these yield increases are attributed to increased use of fertilizers.

TABLE 2

Production and energy requirements per acre of maize in U.S.A. (in 1000 kilocalories).
From Pimentel *et al.* (1973)

Year:	1945	1950	1954	1959	1964	1970
Total energy input:	925	1206	1540	1889	2241	2896
Yield:	3427	3830	4132	5443	6854	8164
Yield/calory input:	3.70	3.18	2.67	2.88	3.06	2.82
Rise in yield:	1909 to 1945 = 26 to 34 bushels/acre.					
	1945 to 1970 = 34 to 81 bushels/acre.					

Examples of other dramatic yield increases are given in Table 3.

An FAO report shows that the use of fertilizers rose considerably between 1961 and 1973 (Table 4). According to a German estimate only 45% of applied nitrogen goes into plant biomass.

According to an FAO report 2.3 milliard tonnes of wood were harvested in the world in 1972; 65% of this was used for fuel, 85% in underdeveloped countries. In Brazil wood is used as fuel for steel making. Total world reserves of wood are estimated at 41 million km^2, of which 50–100,000 km^2, equivalent to an area of one third the Federal German Republic, is cut annually.

An FAO forecast predicts that by the year 2000, 20% of present agricultural land will have become unusable owing to desertification, erosion and salinity. Such losses will be greatest in already over-populated areas. The southern Sahara grows by 10,000km^2 yearly and has advanced 300km. In Italy 50,000km^2 of useful agricultural land was lost by erosion between 1914 and 1934, and in Britain 60,000ha is lost annually to urban and industrial development. 70% of this land is above average quality. "A Blueprint for Survival" (Goldsmith *et al.*, 1972) estimates changes in soil quality worldwide, as shown in Table 5.

Although the exactitude of these figures may be questioned, there is little doubt that they accurately represent current trends. These trends are towards a continued rise in production per unit of land, by means of the increased use of mineral fertilizers and chemicals for pest and weed control. If such increases in yield are at the expense of energy efficiency, i.e. if the ratio of total energy input (in terms of oil and coal, non-renewable resources) to yield is truly decreasing, then present trends probably cannot continue indefinitely.

TABLE 3

Examples of large production increases during the last 15 years.
(From Rehm, 1983)

Product	Country	Period	Production (million tonnes)
Wheat	India	1965	12.3
		1981	36.4
Rice	India	1965	46.0
		1981	82.0
Soja	Brazil	1970	1.5
		1975	10.0
	Argentina	1975	0.9
		1981	3.8
Palm oil	Malaysia	1970	0.5
		1981	2.8
Cassava	Thailand	1967	2.0
		1981	17.9
Cocoa	Ivory Coast	1970	0.2
		1981	0.43
	Brazil	1970	0.18
		1981	0.35
Pineapple	Hawaii	1976	0.5
		1981	1.8
	Philippines	1976	0.4
		1981	1.2
Sunflower	U.S.A.	1976	0.4
		1980	2.0

TABLE 4

World fertilizer use in millions of tonnes.
(FAO Report)

	1949	1961	1973
Nitrogen	3.1	10.2	36.0
P_2O_5	5.0	9.8	22.8
K_2O	3.3	8.5	18.7

TABLE 5

Changes in soil quality worldwide.
(From Goldsmith *et al.*, 1972)

	1882 %	1952 %
Good	85	41
Humus layer half used up	10	38
Exhausted and lost soils	5	20

Shortly after the war there were many predictions that the continued and alarming increase in world population could not be matched by increased food production and that people would starve. People do starve, but for economic rather than scarcity reasons, and in developed countries there are unsaleable surpluses. Prof. Rehm in his lecture addresses himself to the problem of what should be the agricultural strategy to produce adequate food for the 87% of the world population that will shortly be living in developing countries. He considers two strategies: increased subsistence farming over an extending area, including more marginal land where, for climatic and edaphic reasons, production cannot reach above 20–40% of maximum yield, or more intensive use of land suitable (40–80% of maximum yield) or very suitable (80–100% maximum yield) for agriculture. The objective of this talk is to try and show how an understanding, and possible practical exploitation, of mycorrhizal systems can affect our attitude to such problems.

MYCORRHIZA

In 1984 the 6th North American Conference on Mycorrhiza celebrated the centenary of the beginning of mycorrhizal studies and the coining of the term 'mycorrhiza', meaning fungus root. This was applied to the morphologically modified short roots of temperate forest trees, but was soon extended to other types of non-pathogenic root infections that were common in many other plants. By far the most widespread type are the vesicular-arbuscular (VA) mycorrhiza. The soil-based mycelium of these fungi is widely believed to be the largest component of the total fungal biomass in the soil. Although this was already well established by the turn of the century, meaningful studies of VA mycorrhiza only date from about 1960 when it became possible to produce these infections under controlled conditions. This long delay was because the fungi are obligate symbionts and still cannot be grown without a host plant. After only 25 years of experimental study, much remains unknown or at best partly understood.

Mycorrhiza and Plant Nutrition

Phosphate

From an agronomic point of view the most important function of both ecto- and VA mycorrhiza is undoubtedly their ability to increase P supplies to the plant. Because P uptake at the root surface is faster than movement of phosphate to the surface, a depletion zone develops around the root. Hyphae extending from the root surface can pass across this depletion zone and take up phosphate from more distant parts of the soil inaccessible to nonmycorrhizal roots. Within the hyphae phosphate is translocated from the soil-based to the root-based mycelium where the plant provides a readily accessible carbon source for active fungal growth. Neither the site nor the mechanism of phosphate transfer from fungus to root are fully understood. Figure 2 is a diagramatic illustration of the system. The extra phosphate taken up comes from the pool of available soil phosphate, not from the much larger reserves of unavailable, insoluble or strongly adsorbed, soil phosphate. This conclusion, based on ^{32}P labelling experiments, has been widely accepted during the last ten years (see reviews by Mosse, 1978; Tinker, 1978; Hayman, 1982a; Abbott & Robson, 1984). Very recently, however, in a paper confirming earlier experimental results, Bolan *et al.* (1984a) have cast some doubt on the interpretation of such data, particularly in relation to iron phosphates. If their conclusions become widely accepted, this would have interesting implications, particularly for the role of mycorrhiza in lateritic soils. Once inside the hyphal system, the absorbed P is protected against

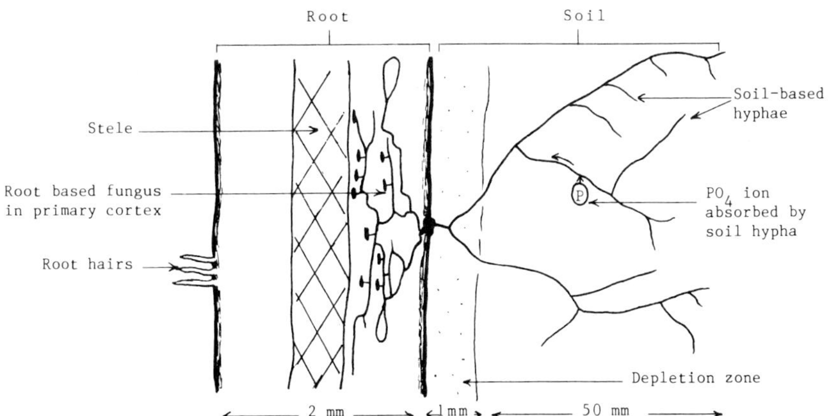

FIGURE 2   Diagram illustrating the mechanism of mycorrhizal uptake of phosphate. (Adapted from Hayman, 1982b).

further immobilisation in the soil. For this reason mycorrhiza often promote a more efficient use of applied fertilizer.

Figure 3 shows a typical response curve of mycorrhizal and non-mycorrhizal plants to applied P. The mycorrhizal plants reach 90% of maximum production at approximately 60 ppm P, whereas the non-mycorrhizal plants require approximately twice as much P. Of particular interest in the present context is the effect of P on infection at both extremes of the curve. Figure 4 shows that with very little phosphate mycorrhizal infection in subterranean clover was also very low, rather more so for an introduced fungus (Bolan *et al.*, 1984b). For both the introduced and indigenous fungus, adding P led to levels supra-optimal for infection, but note that the indigenous fungus was far more sensitive to such additions. With the indigenous fungus maximum infection occurred at a P level equivalent to 6% of maximum shoot growth, but for the introduced fungus only at 62% of maximum growth. In any given system percentage infection is usually related to growth response. It is therefore possible to change responses to applied P by changing the endophyte population. The situation illustrated at the low end of the P response curve may occur in subsistence agriculture, particularly when a more efficient endophyte is introduced. Therefore some fertilizer input is necessary under such conditions to obtain maximum benefit from the introduced endophyte.

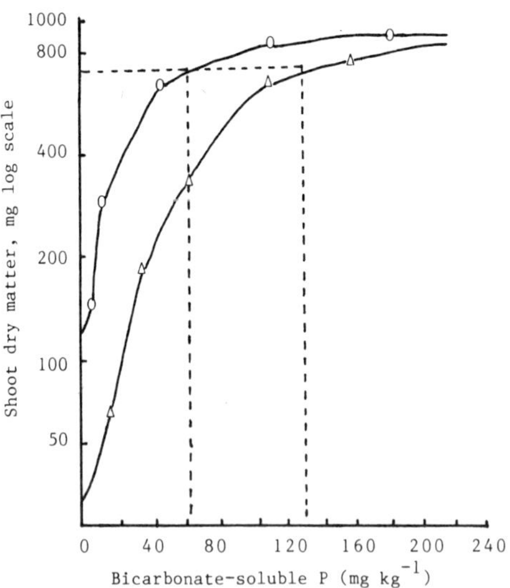

FIGURE 3   P response curve of mycorrhizal and non-mycorrhizal leeks (△—non-mycorrhizal; 0—mycorrhizal). (From Stribley *et al.*, 1980).

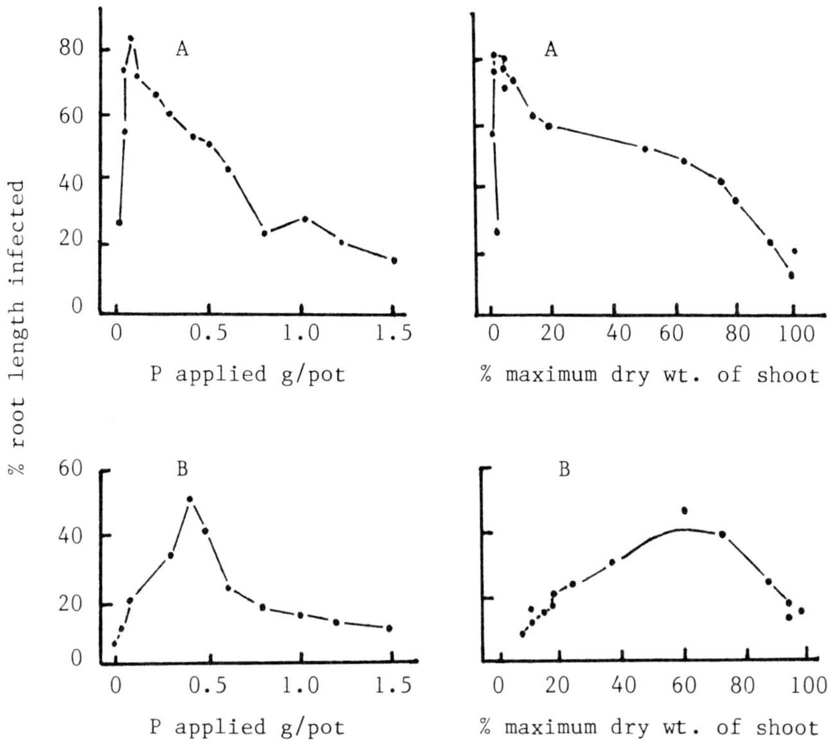

FIGURE 4 Relationship of infection to applied phosphate and yield. A—indigenous fungus; B—introduced fungus. (After Bolan *et al.*, 1984b).

On the other hand, with high P inputs mycorrhizal infection becomes negligible and this may be the case in highly intensive agriculture. Then to obtain the last 5–10% of production from virtually nonmycorrhizal plants requires an inordinately high fertilizer input. Should we therefore content ourselves with 90% of maximum production at 50% the fertilizer input, making full use of a mycorrhizal system?

The exact mechanism of mycorrhizal suppression at high levels of P is not fully understood, but it seems to be related to high P concentration within the plant. Graham *et al.* (1981) attributed the decline in infection to greater membrane stability in, and less leakage and exudation from, P sufficient plants. This would reduce fungal substrate in the rhizosphere and perhaps reduce endophyte development in that region. However, the situation appears to be rather more complex than that. Hayman (1970, 1975) noted that, in the field, added nitrogen can be almost more inhibitory to infection than added P.

This is now a well established finding and applies also to mycorrhiza in *Convulvulus* root organ cultures (Mugnier, personal communication). Whether organic forms of nitrogen are equally inhibitory is not known. In pots, infection can occur even at high P levels provided N is low, or conversely at high N levels provided P is low. In most agricultural situations, however, infection is more likely to be depressed by too much N, and thus one enters as escalating cycle of more nitrogen, less mycorrhiza, more P needed to make the added N effective.

Plants differ greatly in their need for and response to phosphate and in their mycorrhizal dependence. Many temperate Gramineae, because of their extensive systems of fine absorbing roots, only benefit from mycorrhiza in very low P soils. In fact, at higher P levels, they may be better without mycorrhiza. Other plants, however, like cassava or citrus, will never achieve maximum growth without mycorrhizal infection. Table 6 gives some examples in the different dependency groups.

TABLE 6

Mycorrhiza dependency of some crop plants

Strong:	Medium:	Weak:
Cassava	Soja	Wheat
Citrus	Maize	Barley
Onion	Sorghum	Potato
Leek	Paspalum	Rice
Cowpea		
Asparagus		
Stylosanthes guyanensis		
Pinus spp. (ectomycorrhizal).		

Among the most extreme are many temperate forest trees, especially *Pinus* spp., that simply do not establish without mycorrhiza. Evidence is accumulating that even varieties of the same species, for instance wheat (Bertheau *et al.*, 1980; Azcon & O'Campo, 1981) or millet (Krishna *et al.*, 1985), can differ greatly in mycorrhiza dependence.

Many legumes depend strongly on mycorrhiza to obtain the P needed for nodulation and nitrogen fixation (see Munns & Mosse, 1978). Nonmycorrhizal leafless peas inoculated with rhizobia failed to nodulate in a soil containing 20 ppm $NaHCO_3$–soluble P. Some results of an experiment to improve establishment of clover in a Welsh hillside pasture are given in Table 7. There are many situations in the tropics where no nitrogen fixation would occur at all without mycorrhiza.

TABLE 7

Effects of inoculation with an effective mycorrhizal fungus on the
growth of white clover in a Welsh hillside pasture.
(From Hayman & Mosse, 1979)

| Treatment | Dry weight (mg). | | | |
| | Shoot | | Nodules | |
	Inoculated	Control	Inoculated	Control
No P	121	90	1.1	0.1
+ 22 kg P/ha	204	99	2.3	0.2
+ 90 kg P/ha	599	238	9.3	1.6

In several experiments in tropical soils (Mosse, 1977) legumes hardly responded to the addition of rock phosphate unless they were mycorrhizal; Table 8 shows the results of inoculating *Stylosanthes guyanensis* in such a soil from Nigeria.

Almost mystical powers are sometimes claimed for mycorrhiza in relation to composts and organic fertilizers. From 1900 onwards researchers have tried to demonstrate a relationship between soil organic matter and mycorrhizal function. One early hypothesis was that the fungi utilised complex nitrogenous "albuminous" substances in the organic matter and thus improved nitrogen supply to the plant. This may indeed be true for ericaceous mycorrhiza (Stribley & Read, 1975). Later the strong belief in the value of composts was rationalised on the basis that they promoted mycorrhizal infection (Rayner, 1938 & 1939; Howard, 1940). There is relatively little factual evidence to support such claims, but some observations are still being made (Azcon *et al.*, 1978; Nappi *et al.*, 1980). Many ecological surveys have failed to show a clear relationship between soil organic matter and the incidence of mycorrhiza. Peyronel (1943) confirmed his father's observation

TABLE 8

Response of *Stylosanthes guyanensis* to mycorrhizal inoculation and to
rock phosphate in a natural soil from Nigeria (From Mosse, 1977)

Treatment:	Nil	Plus rock phosphate	Mycorrhizal inoculation	Inoculation plus rock phosphate
Dry weight (g)	0.14	0.20	0.65	1.12
Total P (mg)	0.11	0.21	0.79	1.21
N_2 fixation (μ moles C_2H_4/pot/hr)	nil	nil	0.37	0.87

All treatments were inoculated with rhizobium

that in high humus soils the incidence of mycorrhiza was less light dependent than in low humus soils. In citrus (Reed & Frémont, 1935) and in strawberries (Mosse 1959), plants given FYM formed more arbuscules than those given equivalent amounts of mineral fertilizer, and also had more connecting points between the internal and external mycelium. If arbuscules are the main sites of nutrient transfer from fungus to plant, this may be of interest, but the premise is conjectural. It is very probable that the slow nutrient release from organic fertilizers and composts might avoid the rapid rise in plant nutrient concentrations that depresses mycorrhizal development, and that the mycorrhiza benefit nutrient uptake from such slowly available sources in the same way that they improve uptake from rock phosphate. Also the effect of mulches, which improve the moisture holding capacity of the surface soil, would benefit mycorrhizal development, because in many plants there is an important, strongly developed system of mycorrhizal surface roots that may dry out or be destroyed by surface cultivation. Minimal cultivation also would favour these roots which enable plants to exploit the relatively rich surface layers. At the same time mycorrhiza require aerobic conditions. Some effects of mulches (Nappi *et al.*, 1980), of sewage sludge (Spitko & Manning, 1981) and of agronomic practices (Hayman, 1975; Kruckelmann, 1975; Ruissen 1981) have been described, but where results have been expressed in terms of spore numbers they are difficult to evaluate. Both monoculture (Strzemska, 1975) and crop rotations (Hayman, 1982b) can affect the incidence of infection, but results are unpredictable and vary much with crop and fertilizer treatment (Table 9).

TABLE 9

Infection rating after five years of monoculture at two fertilizer regimes. Infection rating is based on examination of root transverse sections (From Strzemska, 1975)

Crop	Year:	Fertilizer regimes			
		I		II	
		1	5	1	5
Rye		173	78	46	11
Wheat		175	269	115	103
Barley		248	222	220	66
Oats		25	176	48	54
Beans*		194	196	116	205

Fertilizer regimes: I=N/P/K kg/ha 60:26:50
        *20:35:83
      II=N/P/K kg/ha 120:52:100
        *40:70:165

Secondary effects of P sufficiency

The most important secondary effect of phosphorus is undoubtedly the better drought resistance of P sufficient plants. Nelsen & Safir (1982) considered that the more rapid recovery from water stress of onions that were mycorrhizal was attributable to the higher P content of mycorrhizal plants from which various other changes in water relations also resulted. Similar conclusions were reached by Hays *et al.* (1984) for *Bouteloua gracilis* (Blue grama) and by Allen & Boosalis (1983) for wheat. It is well documented that mycorrhizal plants transplant better than nonmycorrhizal (Table 10).

TABLE 10

Effects of inoculation on survival of Avocado transplants
(From Menge *et al.*, 1978)

	Mean dry weight (g)	% Healthy plants	Wilt index*
Non-mycorrhizal	17.2	20	2.6
Mycorrhizal	31.4	80	0.4

*Wilt index 0–4

Similar results have been obtained with coffee seedlings in Brazil. Many plantation crops are raised as seedlings in sterilised soils with high mineral fertilizer additions. Such seedlings are therefore nonmycorrhizal and accustomed to high soil nutrient levels. When outplanted into less fertile soils, they often survive badly, especially in drought or other adverse conditions, and die before they develop a mycorrhizal system. With increasing popularity of aseptic clonal propagation the problem of outplanting nonmycorrhizal stock may become more prevalent. For instance Unilevers found that field establishment of their clonally propagated oil palms was poor until they added an inoculum of natural soil before outplanting. Similar problems could arise with asparagus, another highly mycorrhiza-dependent plant now frequently propagated clonally.

Incidence of disease can also be affected by changes in plant nutrient composition resulting from mycorrhizal infection (see below).

Other nutrients

Uptake of minor elements with slow soil diffusion rates, like copper and zinc can also be increased by mycorrhiza (see Mosse, 1973). Uptake of boron was also increased in sunflowers (Jodice *et al.*, 1980).

Recently Buwalda *et al.* (1983) reported a higher concentration of chlorine and bromine in mycorrhizal wheat and barley. They speculated that this might be an intrinsic feature of mycorrhizal systems needing to balance the electro-potential of membranes against the outflow of phosphate into the plant. If indeed this is true, then problems of salinity might be aggravated. One mycorrhizal fungus, however, improved salt tolerance of onions (Ojala, 1982). The relationship of mycorrhizal infection to salt tolerance of cotton (Hamza, 1981) and various other tropical crops (El Deepah, 1981) has been the subject of two theses.

Mycorrhiza may be involved to some extent in the problem of acid rain. Dutch research workers believe that, where trees are showing signs of damage, their mycorrhiza are also impaired, whereas isolated groups of undamaged trees appear to have well developed mycorrhizal roots. One is faced here with the age old problem of cause and effect which may be difficult to resolve. Killham & Firestone (1983) observed that mycorrhiza increased susceptibility of a perennial grass to simulated acid rain. In *Calluna*, however, mycorrhizal infection protected plants against toxicity of copper and zinc, and led to a reduction in heavy metal content of shoots (Bradley *et al.*, 1981). *In vitro*, germination and growth of germ tubes of some VA fungi are very sensitive to quite low concentrations of zinc and manganese (Hepper & Smith, 1976).

Fabig (1982) examined the effects of heavy metals and of aluminium, on mycorrhizal responses of some tropical and subtropical plants. Gildon & Tinker (1983) found that heavy metals above normal soil concentrations also inhibit mycorrhiza. However, in one, naturally occuring, cadmium and zinc toxic soil there were well developed mycorrhiza, apparently formed by a specially adapted race of an endophyte normally sensitive to cadmium (Gildon & Tinker, 1981). The question of differential tolerance of different endophytes to low or high soil pH, low or high plant nutrient levels, low or high temperature, and the possibility of adaptive races within species, is so far little documented. It will almost certainly play a role as agriculture spreads into virgin land. For instance it is essential to lime cerrado soils of Brazil before they can be used for agriculture because of their high aluminium content. After liming the indigenous endophyte population is poorly adapted to the new, higher pH. New, better adapted endophytes will develop or arrive by natural processes, but such development may be slow, especially where large areas are concerned. Introducing better adapted endophytes of known efficiency may speed the processes of natural selection.

Mycorrhiza and Biocides

Modern intensive agriculture tends towards monoculture. This raises special problems of disease and weed control which lead to increased use of biocides.

Applied at recommended rates most crop sprays appear to be relatively harmless to mycorrhizal fungi. Some cause temporary reductions in infection. Among the most toxic is benomyl. Some nematicides actually appear to stimulate mycorrhiza. In Hawaian pineapple plantations where nematicides are regularly applied, roots are virtually 100% mycorrhizal and soils contain large and varied spore populations. Saif (1983) lists 129 publications on biological interactions between mycorrhiza and pathogens, including effects on take-all disease of wheat, Verticillium wilt, *Phytopthora* and *Pythium* root rots, nematodes and viruses. Mycorrhiza can reduce disease by producing antibiotics, influencing sporulation pattern of the pathogen, by competing for infection sites and substrates or by affecting nematode life cycles. They can also increase disease tolerance in better nourished plants, but they never provide a complete control of disease. Schenck & Kellam (1978) and Dehne (1982) have reviewed the subject.

Most soil sterilants and fumigants, particularly methyl bromide will also kill mycorrhizal fungi. Citrus or liquidambar planted in such soils will often grow poorly and suffer from minor element deficiencies. These can be overcome by inoculation with mycorrhizal fungi, proving that, in this instance, the lack of mycorrhiza is the cause not the effect of the poor growth. Natural recolonisation of such sites occurs by wind-blown particles, soil fauna or regeneration from deeper soil layers not reached by the sterilant. The mechanisms involved are not well understood.

Mycorrhiza, Erosion and Degraded Soils

I have already touched on the rather tenuous relationship between mycorrhiza and soil organic matter. Tisdall & Oades (1979) reported a much more direct effect of mycorrhiza on soil aggregation (Table 11). Particularly aggregates above 2mm were greater in soils with mycorrhizal plants. This should significantly affect soil stability. It was shown that the soil mycelium of VA fungi was coated with a mucilagenous substance that caused soil particles to adhere.

TABLE 11

Effect of inoculation on water-stable aggregates (From Tisdall & Oades, 1979)

Soil treatment:	Hyphal Length metres/gram soil	% stable aggregates >2mm	Dry weight shoot (g)
Fumigated	5.9	10.2	17.8
Fumigated-inoculated	13.7	15.9	17.8
Non-fumigated	11.3	13.5	18.1

Related to erosion is the problem of degraded soils following deforestation and major soil disturbances that entail removal and disturbance of the top soil. Even where this is done carefully, the stored soil looses infectivity after about a year. When it is replaced, the new vegetation that develops will consist predominantly of nonmycorrhizal plants, often Chenopods. A normal flora re-establishes slowly unless inoculation is practised. Problems of VA mycorrhizae and reclamation of arid and semi-arid lands were discussed at a conference in Wyoming (Williams & Allen, 1984).

The most talked about de-forestation is that which has occurred in Amazonia in Brazil. Once the luxurious vegetation is destroyed, the remaining soil is found to be disastrously P deficient so that very little will grow on it once the humus layer has gone. Mycorrhiza are supposed to play an important role in tropical forests in re-cycling nutrients directly from the litter layer, thus preventing their loss by leaching in torrential rains. Most of the phosphate, mined during decades of forest growth, is re-cycled in the standing crop, and lost when this is removed. There is also some evidence that mycorrhiza may form organic links within which nutrients can move from one plant species to another (Whittingham & Read, 1982).

CONCLUSIONS

I would like to make two points in conclusion.

1) I have not discussed at all the practical possibilities of utilizing mycorrhizal systems for modifying methods of production so as to avoid the dangers of over-exploitation. We know that we can re-establish mycorrhizal fungi by inoculation where they have been eliminated from soils. We can speed up mycorrhizal development in young seedlings of annual plants by inoculation or by growing a nurse crop that will build up the indigenous population of mycorrhizal fungi in a soil. We can also introduce new, more effective mycorrhizal fungi to replace a less effective indigenous population. All this has been done experimentally under field conditions. What is in question is the cost. Because we cannot culture the fungi, inoculum has to be produced in conjunction with a host plant and it is bulky. Nevertheless several commercial companies are interested today in producing VAM inocula. Ectomycorrhizal studies do not suffer from this limitation. Many of the fungi can be grown in fermentors and produce good inocula with a reasonable shelf life when embedded in polyacrylamide gels (LeTacon et al., 1983).

2) It is as normal for the roots of plants to be mycorrhizal as it is for their leaves to photosynthesise. Any agricultural operation that disturbs the natural ecosystem will have repercussions on the mycorrhizal system. This is complex consisting essentially of three components, the plant, the fungus and the soil. There is little host specificity, but plant species responses to infection

vary greatly. While we can, perhaps, claim a working knowledge of the potential of mycorrhiza for improving plant phosphate nutrition, we know very little indeed about interactions between mycorrhizal fungi and the soil, including its complex microbial population. The selection of suitable mycorrhizal fungi for particular situations, a long term objective of mycorrhizal research, is at present a matter of guesswork.

In evolutionary history mycorrhizal associations are very old. It has been suggested that they evolved simultaneously with the development of land plants. If modern agricultural methods ultimately lead to the disappearance of mycorrhizal associations in our crop plants, the full consequences of this are not easily predictable.

SUMMARY

Some causes for doubt about the sustainability of present day agriculture are briefly discussed. The mechanism and function of vesicular-arbuscular systems are considered in relation to these problems with special reference to plant nutrient uptake, particularly phosphorus, to effects of mineral fertilizers, management practices, salinity, biocides and disease control, problems of erosion and degraded soils. In natural ecosystems most plants form mycorrhizal associations and many agricultural practices will have repercussions on mycorrhizal systems. Because of the range of mycorrhizal fungi, the inadequate understanding of soil: fungus relationships and the variability of plant responses, it is difficult to predict the ultimate effects on crop production of modifications in mycorrhizal systems.

References

Abbott, L.K. & Robson, A.D. (1984). The effect of mycorrhizae on plant growth. In *VA Mycorrhiza* (C. Ll. Powell & D.J. Bagyaraj, eds.), chapter 6. CRC, Uniscience.

Allen, M.F. & Boosalis, M.G. (1983). Comparative water relations of winter wheat infected by two vesicular-arbuscular mycorrhizal fungi. *New Phytologist,* **93**, 61–67.

Azcon, R. & O'Campo, J.A. (1981). Factors affecting the vesicular-arbuscular infection and mycorrhizal dependency of thirteen wheat cultivars *New Phytologist,* **87**, 677–685.

Azcon, R., Barea, J.M. & Montoya, E. (1978). Biological fertilisation with 'VA' mycorrhizae and phosphobacteria. II. Effect of manuring and timing of phosphate applications on mycorrhizal infection of *Lavendula spica* L. in the seedbed. *Anales de edafologia y Agrobiologia,* **37**, 99–104.

Bertheau, Y., Gianinazzi-Pearson, V. & Gianinazzi, S. (1980). Development and expression of endomycorrhizal association in wheat. 1. Evidence of a varietal effect. *Annales de l'amélioration des plantes,* **30**, 67–78.

Bolan, N.S., Robson, A.D., Barrow, N.J. & Aylmore, L.A.G. (1984a). Specific activity of phosphorus in mycorrhizal and non-mycorrhizal plants in relation to the availability of phosphorus to plants, *Soil Biology & Biochemistry,* **16**, 299–304.

Bolan, N.S., Robson, A.D. & Barrow, N.J. (1984b). Increasing phosphorus supply can increase the infection of plant roots by vesicular-arbuscular mycorrhizal fungi. *Soil Biology & Biochemistry,* **16**, 419–420.

Bradley, R., Burt, A.J. & Read, D.J. (1981). Mycorrhizal infection and resistance to heavy metal toxicity in *Calluna vulgaris. Nature,* **292**, 335–337.

Buwalda, J.G., Stribley, D.P. & Tinker, P.B. (1983). Increased uptake of bromide and chloride by plants infected with vesicular-arbuscular mycorrhizas. *New Phytologist,* **93**, 217–225.

Dehne, H.W. (1982). Interactions between vesicular-arbuscular mycorrhizal fungi and plant pathogens. *Phytopathology,* **72**, 1115–1119.

Fagig, B. (1982). Influence of aluminium and the heavy metals Fe, Mn, Zn, Pb and Cd on the development and the efficiency of vesicular-arbuscular mycorrhiza in tropical and subtropical plants. Ph.D. thesis, University of Göttingen. 180 pp.

El Deepah, H.R.A. (1981). Efficiency of vesicular-arbuscular mycorrhiza in seven crop species, modified by NaCl in the soil. Ph.D. thesis, University of Göttingen. 87 pp.

Gildon, A. & Tinker, P.B. (1981). A heavy metal tolerant strain of a mycorrhizal fungus. *Transactions of the British Mycological Society,* **77**, 648–649.

Gildon, A. & Tinker, P.B. (1983). Interaction of vesicular-arbuscular mycorrhizal infection and heavy metals in plants. I. The effects of heavy metals on the development of vesicular-arbuscular mycorrhizas. *New Phytologist,* **95**, 247–262.

Goldsmith, E., Allen, R., Allaby, M., Davoll, J. & Lawrence, S. (1972). *A Blueprint for Survival.* Penguin; Harmondsworth.

Graham, J.H., Leonard, R.T. & Menge, J.A. (1981). Membrane mediated decrease in root exudation responsible for phosphorus inhibition of vesicular-arbuscular mycorrhiza formation. *Plant Physiology,* **68**, 548–552.

Gruhl, H. (1975). *Ein Planet wird geplündert.* Fischer Taschenbuchverlag. 383 pp.

Hamza, E.A. (1981). Effect of vesicular-arbuscular mycorrhiza on cotton under the influence of NaCl in the soil. Ph.D. thesis, University of Göttingen. 95 pp.

Hayman, D.S. (1970). *Endogone* spore numbers in soil and vesicular-arbuscular mycorrhiza in wheat as influenced by season and soil treatment. *Transactions of the British Mycological Society,* **54**, 53–63.

Hayman, D.S. (1975). The occurrence of mycorrhiza in crops as affected by soil fertility. In *Endomycorrhizas* (F.E. Sanders, B. Mosse & P.B. Tinker, eds.), pp. 495–509. Academic Press; London

Hayman, D.S. (1982a). Influence of soils and fertility on activity and survival of vesicular-arbuscular mycorrhizal fungi. *Phytopathology,* **72**, 1119–1125.

Hayman, D.S. (1982b). Practical aspects of vesicular-arbuscular mycorrhiza. In *Advances in Agricultural Microbiology* (N.S. Subba Rao, ed.), pp. 325–373. Oxford and I.B.H. Publishing Co.

Hayman, D.S. & Mosse, B. (1979). Improved growth of white clover in hill grasslands by mycorrhizal inoculation. *Annals of Applied Biology,* **93**, 141–148.

Hays, R., Reid, C.P.P. & Coleman, D.C. (1984). Effect of vesicular-arbuscular mycorrhiza on drought, nutrient uptake and drought recovery in Blue Grama. Abstract in *VA Mycorrhizae and Reclamation of Arid and Semi-arid Lands* (S.E. Williams & M.F. Allen, eds.). University of Wyoming Report No. SA1261.

Hepper, C.M. & Smith, G.A. (1976). Observations on the germination of *Endogone* spores. *Transactions of the British Mycological Society,* **66**, 189–194.

Howard, Sir A. (1940). *An Agricultural Testament.* Oxford University Press. 253 pp.

Jenkinson, D.S. & Johnstone, A.E. (1976). Soil organic matter in the Hoosfield continuous barley experiment. Rothamsted Annual Report for 1976, Part II, pp. 87–101.

Jodice, R., Nappi, P. & Luzzati, A. (1980). Effect of vesicular-arbuscular mycorrhiza and organic matter on boron nutrition of sunflower. *Allionia,* **24**, 43–48.

Killham, K. & Firestone, M.K. (1983). Vesicular-arbuscular mediation of grass response to acidic and heavy metal depositions. *Plant & Soil,* **72**, 39–48.

Krishna, K.R., Shetty, K.J., Dart, P.J. & Andrews, D.J. (1985). Growth and phosphorus uptake responses to mycorrhizal inoculation is plant genotype dependent. Proceedings of the 6th North American Conference on Mycorrhizae, 1984, p.403. Bend; U.S.A.

Kruckelmann, H.W. (1975). Effects of fertilizers, soils, soil tillage, and plant species on the frequency of *Endogone* chlamydospores and mycorrhizal infection in arable soils. In *Endomycorrhizas* (F.E. Sanders, B. Mosse & P.B. Tinker, eds.), pp. 511–525. Academic Press; London.

LeTacon, F., Jung, G., Michelot, P. & Mugnier, J. (1983). Efficacité en pépinière forestière d'un inoculum de champignon ectomycorrhizien produit en fermenteur et inclus dans une matrice de polymère. *Annales des sciences forestières,* **40**, 165–176.

Menge, J.A., Davis, R.M., Johnson, E.L.V. & Zentmyer, G.A. (1978). Mycorrhizal fungi increase growth and reduce transplant injury in avocado. *California Agriculture,* **32**, 6–7.

Mosse, B. (1959). Observations on the extramatrical mycelium of a vesicular-arbuscular endophyte. *Transactions of the British Mycological Society,* **42**, 439–448.

Mosse, B. (1973). Advances in the study of vesicular-arbuscular mycorrhiza. *Annual Review of Phytopathology,* **11**, 171–196.

Mosse, B. (1977). Plant growth responses to vesicular-arbuscular mycorrhiza. X. Responses of *Stylosanthes* and maize to inoculation in unsterile soils. *New Phytologist,* **78**, 277–288.

Mosse, B. (1978). Vesicular-arbuscular mycorrhiza research for tropical agriculture. Research Bulletin 194, Hawaii Institute of Tropical Agriculture and Human Resources, University of Hawaii. 82 pp.

Munns, D.N. & Mosse, B. (1980). Mineral nutrition of legume crops. In *Advances in Legume Science* (R.J. Summerfield & A.H. Bunting, eds.), pp. 115–125. International Legume Conference, Kew, 1978.

Nappi, P., Jodice, R. & Kofler, A. (1980). Vesicular-arbuscular mycorrhiza in vineyards given different soil treatments in Southern Tirol. *Allionia,* **24**, 27–42.

Nelsen, C.E. & Safir, G.R. (1982). Increased drought tolerance of mycorrhizal onion plants caused by improved phosphorus nutrition. *Planta,* **154**, 407–413.

Ojala, J.C. (1982). Soil factors influencing vesicular-arbuscular mycorrhizal plant growth. Ph.D. thesis, University of California; Riverside. 105 pp.

Peyronel, B. (1943). Richerche sulla simbiosi micorrizica nelle Epatiche. *Nuovo giornale botanico italiano e Bolletino della Societa botanica italiana,* **49**, 362–382.

Pimentel, D., Hurd, L.E., Bellotti, A.C., Forster, M.J., Oka, I.N., Sholes, O.D., & Whitman, R.J. (1973). Food production and the energy crisis. *Science,* **182**, 443–448.

Rayner, M.C. (1938). Use of soil or humus inocula in nurseries and plantations. *Empire Forestry Journal,* **17**, 236–243.

Rayner, M.C. (1939). The mycorrhizal habitat in relation to forestry. III. Organic composts and the growth of young trees. *Forestry,* **13**, 19–35.

Reed, H.S. & Frémont, Th. (1935). Étude physiologique de la cellule à micorrhizes dans les racines de Citrus. *Révue cytologique,* **1**, 327–348.

Rehm, S. (1983). Sicherung und Steigerung der pflanzlichen Produktion in Entwicklungsländern. Göttinger Beitr. z. Land-und Forstwirtschaft in den Tropen und Subtropen. Heft 4. 25 pp.

Ruissen, M.A. (1982). The development and significance of vesicular-arbuscular mycorrhizas as influenced by agricultural practices. Ph.D. thesis, Landbouwhogeschool, Wageningen. 111 pp.

Saif, S.R. (1983). *Bibliography on vesicular-arbuscular mycorrhiza 1970–1982.* Centro Internacional de Agricultura Tropical. ISSN 0120-5137. 143 pp.

Schenck, N.C. & Kellam, M.K. (1978). The influence of vesicular-arbuscular mycorrhizae on disease development. Technical Bulletin 798, Florida Agriculture Experiment Station; Gainesville. 16 pp.

Spitko, R.A. & Manning, W.J. (1981). Irradiated digested sewage sludge: effects on plant symbiont associations in the field. *Environmental Pollution A,* **25**, 1–8.

Stribley, D.P., Tinker, P.B. & Snellgrove, R.C. (1980). Effect of vesicular-arbuscular mycorrhizal fungi on the relations of plant growth, internal phosphorus concentration and soil phosphate analysis. *Journal of Soil Science,* **31**, 655–672.

Strzemska, J. (1975). Mycorrhiza in farm crops grown in monoculture. In *Endomycorrhizas* (F.E. Sanders, B. Mosse & P.B. Tinker, eds.), pp. 527–535. Academic Press; London.

Tinker, P.B. (1978). Effects of vesicular-arbuscular mycorrhizas on plant nutrition and growth. *Physiologie vegetale,* **16**, 743–751.

Tisdall, J.M. & Oades, J.M. (1979). Stabilisation of soil aggregates by the root systems of ryegrass. *Australian Journal of Soil Research,* **17**, 429–441.

Whittingham, J. & Read, D.J. (1982). Vesicular-arbuscular mycorrhiza in natural vegetation systems. III. Nutrient transfer between plants with mycorrhizal interconnections. *New Phytologist,* **90**, 277–284.

Williams, E.W. & Allen, M.F. eds. (1984). *VA Mycorrhizae and Reclamation of Arid and Semi-arid Lands.* Wyoming Agricultural Experiment Station Report No. SA1261. 91 pp.

Plant Health and the Sustainability of Agriculture, with Special Reference to Disease Control by Beneficial Microorganisms

R. James Cook

United States Department of Agriculture, Agricultural Research Service, Pullman, WA, 99164, U.S.A.

INTRODUCTION

Since the first civilizations, man has found it increasingly more important to pay attention to the health of crop plants. However, never before has the green plant and its health been more important to our future well-being. Browning (1983) suggests that the 1973 world energy crisis ushered in the "Age of Plants." The supply of solar biomass preserved as oil and which has contributed so significantly to the expansion of civilization and international commerce will eventually become exhausted. More and more, nations will become dependent on their annual production of solar biomass. We must learn to provide for our needs each year on what we can grow, and we must find ways to achieve the yield potential of our crops while conserving our resources.

Considering the critical importance of green plants to our survival now and into the future, it is axiomatic that these plants must be healthy. Unfortunately plant health is usually taken for granted except when obviously damaged by some pest or abiotic stress, or killed prematurely by an epidemic disease. In too many cases, the effects of less obvious plant infections are blamed on the weather or inadequate soil fertility. Imagine, by analogy, if our understanding of human health was still at a stage where the flu was attributed to cold, wet weather instead of a virus. This paper, therefore, begins with a brief discussion of the importance of plant-disease control to a sustainable agriculture.

PLANT-DISEASE CONTROL AND THE SUSTAINABILITY OF AGRICULTURE

A healthy crop is more than a crop without obvious leaf spots, rust, smut, or blight; although certainly these important maladies have caused countless crop failures and mass starvation at times in the past. A healthy crop is free of both major and minor parasitic diseases—the diseases we can see and the diseases that often we cannot see but which prevent the crop from growing and yielding at its potential. A world record yield of wheat was produced in 1961 in Grant County, Washington State at 209 bushels/acre (14 mt/ha), on land that was virgin desert (virtually free of wheat pathogens) only a few years earlier and in a year when rust and other foliar diseases were insignificant in the area. The same inputs of irrigation water and fertilization today result in yields of only half (on average) of this record yield. A situation whereby the inputs are adequate for 14 mt/ha but the yield achieved is only 7–8 mt/ha obviously is not compatible with a sustainable agriculture. Unfortunately, the average yield for a given crop in a given area and field is commonly only 50–60% of the best yield obtained for that crop in that area and year and with basically the same weather and agronomic inputs (Cook, 1985b).

There is a potential yield for every crop, field, and year that is set by the climate, weather, edaphic conditions, agronomic inputs, and genetics of the cultivar. This is the "biological potential" (Cook & Baker, 1983) or attainable yield (Zadocks & Shein, 1979). The actual yield of any given crop, field, and year is a reflection of weather, edaphic conditions, genetics of the cultivar, and agronomic inputs (fertilizers, irrigation water, tillage, etc.), counterbalanced by the constraints imposed by diseases, insect pests, nematodes, and competition from weeds. The agronomic inputs are necessary to achieve the biological potential, and in an ideal year with irrigation and adequate fertilization, the biological potential may approach the absolute potential or theoretical maximum. This is probably what happened in 1961 when one grower with irrigated wheat land produced 14 mt/ha. The constraints of diseases and pests, on the other hand, limit the actual yield to some level less than the biological potential, sometimes only 50% or even less of the biological potential (Cook, 1985b). It is possible to override the effects of some pests or diseases, e.g. certain root diseases, by adding more fertilizer and/or water, but it is probably more efficient and certainly more compatible with a sustainable agriculture to control or eliminate the pests or diseases to the extent possible and permit the plant to grow more nearly at its potential.

There is a tendency to believe that the exceptionally high crop yields are not sustainable ecologically and that lower yields are more optimal and hence more compatible with a sustainable agriculture. This concept may not be valid. Some exceptionally high yields of wheat and rice are being achieved in the People's Republic of China using surface water stored in reservoirs for irrigation and by fertilizing with recycled organic wastes. The inputs as human labor are enormous, but there is no suggestion that the method is not sustainable ecologically. Part of their success in maintaining high yields may be that their fertilizer, being largely an organic source, provides not only plant nutrients but also contributes to better root health by improving soil structure and suppressing or eliminating inoculum of soilborne plant pathogens (Kelman & Cook, 1977; Williams, 1979). Improvements in root health, by making the root system more efficient, can contribute as much if not more to growth and yield of a crop as can high rates of fertilizers (Wilhelm & Paulus, 1980; Cook, 1984a; 1984b). The difficulty arises when the grower increases the fertilizer and irrigation inputs to obtain a yield he knows is possible, but still the yield is disappointingly low because of constraints from diseases, insects, nematodes, weeds and abiotic stresses. Applying the water and nutrients necessary for a truly healthy crop to achieve a yield at or near its biological potential is not necessarily incompatible with a sustainable agriculture.

CAUSES OF PLANT DISEASE

Plant disease agents may be grouped into two categories: (1) the well-known parasitic fungi, bacteria, viruses, viroids, and nematodes, and (2) a lesser-known group of nonparasitic pathogens. The latter, sometimes referred to as minor pathogens, subclinical pathogens, deleterious microorganisms, and exopathogens (Cook & Baker, 1983), includes a select group of mainly epiphytes that cause direct damage to plant tissues or growth disturbances by secretion of toxic substances or growth regulators. Some of these plant colonists may accelerate senescence, and strains of *Pseudomonas syringae* act as ice-nucleation agents to initiate frost damage (Lindow, 1983). Disease agents limit yield by themselves, but more commonly they interact with abiotic factors to produce greater damage than either one could produce alone. For example, poor physical conditions of the soil commonly predisposes roots to damage caused by pathogens, and infected roots are often less able to cope with poor physical conditions of the soil. Plants with vigor limited by pathogens are less competitive with weeds, and plants unable to take up adequate nutrients because of root disease may be more susceptible to weak parasites.

Of all constraints to plant health, none are more critical or overlooked more frequently than are the biotic constraints to root health (Wilhelm & Paulus, 1980; Cook, 1984a). Root pathogens can destroy an entire root sysem, but subtle damage such as accelerated death of rootlets, loss of root hairs, and cessation of growth of root tips can be just as important. The production capability of a crop with a fully healthy root system can be estimated experimentally using soil fumigation as a research tool. It is also possible by use of this tool to separate the biotic from the abiotic constraints to root health. Many workers have reported in years past on the often spectacular increased growth response of plants to soil fumigation (Wilhelm, 1965), but explaining the effect has been controversial because there is a flush of nitrogen and other mineral nutrients that complicates interpretations. For crops studied in any detail, the results are now clear that the response is due to improvements in root health because of the elimination of pathogens (Wilhelm & Paulus, 1980; Cook, 1984a).

In eastern Washington, rainfed wheat (annual precipitation about 50 cm) yields 1,000–2,000 kg/ha (15–25%) more when grown in soil treated with a soil fumigant to eliminate inoculum of *Pythium* spp. (Chamswarng & Cook, 1985) before sowing than does wheat grown in adjacent non-fumigated plots (Cook & Haglund, 1982). The fumigation treatments may include chloropicrin, a 1,3-dichloropropene chloropicrin mixture (Telone C), or methyl bromide. The flush of nitrogen was ruled out experimentally as an explanation for the effect on the basis that 1,3-dichloropropene (Telone) alone and Telone-C both produced a flush (10–20 kg/ha) of mineral nitrogen, and the nitrogen remained for most of the growing season as ammonium, but only the latter eliminated *Pythium* inoculum and gave an increased growth response (Cook & Haglund, 1982). Wheat grown in fumigated plots also survives better during severe winters and begins growing sooner in the spring (unpublished), suggesting that root pathogens, mainly *Pythium* spp., predispose the wheat to winter injury.

Pathogen inoculum is also eliminated from soil by the heat produced by burning crop residue. In plots at Pullman, about 75% of the *Pythium* inoculum was eliminated from the top 5 cm of soil by burning a layer of fresh, bright wheat straw placed on the soil surface. The layer was about 30 cm thick (uncompacted), or approximately the amount left as a wind-row if not spread by the harvester. The yield was about 20% (1,050 kg/ha) greater where burned, about the same as where solarized (unpublished data). Ash alone gave no yield response. The intent is not to recommend burning but rather, to reinforce the point scientifically that root disease caused by identifiable agents and not the lack of fertility or water was the limiting factor to wheat growth in this traditional wheat-producing area.

Some discount these results on the basis that farmers cannot afford to fumigate and should not burn their wheat fields, but this argument misses the point that existing cultivars are capable of yielding 15–25% more with no more fertilizer or water, but are prevented from doing so because of root pathogens. Moreover, this evidence for the significance of root pathogens as common and widespread constraints to crop productivity must be kept in mind when attempting to explain the effects of practices such as crop rotation, organic amendments, liming, and many others on plant growth and yield. Practices that favor root disease, e.g. monoculture and reduced tillage, will result in lower yields and those that eliminate or suppress pathogens, e.g. crop rotation, organic amendments, and burning will result in greater yields.

To reveal the full production capability of a crop as set by the local weather, climate, soil conditions, available water, and agronomic inputs can be done only by satisfying at least three critiera. 1) The planting material must be pathogen-free, i.e. not infected with some pathogenic virus, fungus, or bacterium. For vegetatively-propagated plants such as potatoes (Slack, 1980) and bananas (Hwang *et al.*, 1984), this has required the culture of pathogen-free plantlets from meristems. 2) The planting material must be grown in pathogen-free soil. The soil need not be sterile, only pathogen free. This criterion is the most difficult to satisfy and is most commonly ignored. 3) The foliage must be protected all season from foliar disease and insect-vectored viruses. If any one of these criteria are not satisfied, the plant will not grow at its potential. It may be too expensive commercially to satisfy all three criteria, but at least one has a better understanding for why the level of productivity is less than expected or why it varies with cultural practices or from area to area within fields. Our challenge is how to achieve disease control using beneficial microorganisms or other biological approaches as a supplement to or instead of soil fumigation and the foliar pesticides.

THE USE OF MICROORGANISMS TO PROTECT PLANT HEALTH

A great deal has been written on the use of microorganisms for biological control of plant pathogens, including two books (Baker & Cook, 1974; Cook & Baker, 1983) and several proceedings of symposia (Baker & Snyder, 1965; Toussoun *et al.*, 1972; Bruehl, 1975; Schippers & Gams, 1979; Papavizas, 1981). Since the present symposium is on the use of microorganisms in a sustainable agriculture, the balance of this paper will focus on their use to control plant diseases. The subject is presented in two parts: The use of microorganisms to eliminate pathogen inoculum and their use to limit pathogenesis.

The Use of Microorganisms to Eliminate Pathogen Inoculum from Soil

There are at least five practices or treatments that promote biological destruction of pathogen inoculum in soil. In all cases the destruction results partially or entirely from resident (naturally occuring) antagonistic micro-organisms. The practices or treatments are: crop rotation; tillage; organic amendments; soil flooding; and solarization. With the exception of solarization, all are old and familiar but are still "the state of the art" so far as biological control of pathogen inoculum in soil is concerned.

Crop rotation. Crop rotation provides a means whereby more time is made available for naturally biological destruction of pathogen inoculum in soil. Monoculture selects in favor of populations of soil microorganisms adapted to the roots of the crop grown in monoculture. Among these microorganisms, some may have a deleterious effect on the roots (exopathogens *sensu* Cook & Baker, 1983), by their liberation of toxic or growth-regulating substances, and a few, possibly as few as only one and probably no more than five or six, will be parasitic on the roots. Because of specialization, those parasitic on one crop usually can be controlled by growing an unrelated crop. Without recourse to a living host, the specialized root parasites gradually starve and are eliminated from soil by the microflora and fauna (Garrett, 1956).

Earlier theory held that yields go down with monoculture because the soil is depleted of nutrients, but it is more likely that the yield decline results from increased damage by pathogens specialized for the monoculture crop. Crop rotation is sometimes as effective as soil fumigation for control of soilborne plant pathogens. Moreover, like the effects of soil fumigation, the healthier appearance of the crop can give the impression that soil fertility has been improved, but actually it is root health and thus uptake of plant nutrients that has been improved. Some major differences between crop rotation and soil fumigation are that the destruction is mainly biological rather than chemical, no biological vacuum is left by the treatment, but it takes up to 2–3 years to achieve the effect. An important advantage of rotation over fumigation is that a more resilient microbiological balance is left to limit recolonization of the soil by root pathogens when a susceptible crop is again grown in the field.

Tillage. Many plant pathogens- root as well as leaf-infecting- exist between host crops in the residue of their host colonized earlier through pathogenesis. This residue then serves as a foodbase and springboard for the pathogen to infect the next crop (Cook *et al.*, 1978). Biological destruction of pathogen inoculum in crop residue can be accelerated in many cases by frequent stirring of the soil, probably because the temporary burst of microbial activity increases the stress placed on the pathogen by the associated microbiota, and also because new surfaces of the infested residue become exposed for colonization by saprophytes. As with the use of crop rotation, the more vigorous growth of a crop in well-tilled soil can give the impression that fertility has been improved, and sometimes this is the case, but root health and hence probably the efficiency of nutrient uptake is also improved owing to a lower inoculum potential of one or more root pathogens.

In the Pacific Northwest, take-all caused by *Gaeumannomyces graminis* var. *tritici* (Moore & Cook, 1984), Cephalosporum stripe caused by *C. gramineum* (Latin *et al.*, 1982), bare patch caused by *Rhizoctonia solani* (Weller *et al.*, 1985), and root damage caused by *Pythium* spp. (Cook *et al.*, 1980) all are more important on wheat where the residue is left near the soil surface and the soil is disturbed only minimally (as with some forms of conservation tillage) than where conventional clean tillage is used. Pythium root rot, in particular, can be very severe, causing acute stunting of wheat, loss of stand during the winter months, and nutrient deficiency-like symptoms in the tops (Cook & Haglund, 1982). The field symptoms commonly attributed to phytotoxins from rotting straw (Elliott *et al.*, 1978) and nitrogen "tie-up" (Power & Legg, 1978) may in many if not most instances be caused by this complex of *Pythium* spp. The disease-control benefits of tillage were not recognized until minimum- and no-tillage became popular. Unfortunately, intensive tillage is not compatible with a sustainable agriculture, mainly because of the greater soil loss from erosion. The problem has been compounded by the fact that wheat growers have also been tending toward shorter or no crop rotations with wheat. Intensive tillage can be used as a substitute for some crop rotation, but it is not possible at present to maintain adequate root health for wheat, or for many other crops for that matter, with neither crop rotation nor tillage and without soil fumigation.

Organic Amendments. The incorporation of large quantities of organic matter into soil can have an eradicative effect on inoculum of soilborne plant pathogens (Baker & Cook, 1974; Papavizas & Lumsden, 1982), but there are exceptions (Kaiser & Horner, 1980). The first report of the use of soil microorganisms to control a pathogen was through a stimulation of antagonism to the potato scab organism by the addition of organic materials (Sanford, 1926; Milard & Taylor, 1927). Many examples have emerged since these early reports. In Taiwan, a special blend of organic and inorganic materials and referred to as the SH mixture controls fusarium wilts, root rots, and several other diseases caused by soilborne pathogens (Sun & Huang, 1985). In Australia, incorporation of large quantities of organic materials around the roots of avocado trees, so as to maintain the organic matter of soil at a content near that of the surrounding undisturbed rainforest, provides highly effective control of phytophthora root rot of the avocado (Broadbent & Baker, 1975). In Texas, organics applied prior to planting cotton have long been known to control phymatotrichum root rot of cotton (King *et al.*, 1934). The microbial activity stimulated by the added energy supply results in more intense forms of biological stress on the propagules of pathogens. Conceivably, propagules are temporarily predisposed to attack by anaerobes during periods of intense microbial activity, when large volumes of soil are likely to be anaerobic (Cook & Baker, 1983). Propagules are also stimulated to germinate by solutes and/or volatiles released from the organic matter, but having no host to infect, the germlings then die and are destroyed (Papavizas & Lumsden, 1982). Populations of saprophagus nematodes may increase, which then favor multiplication of nematode trappers that lower the population of both plant parasitic and nonparasitic nematodes. However, if the material is very fresh, e.g. fresh cow manure, *Pythium* spp. can be stimulated to cause greater damage (Kaiser & Horner, 1980). Organic amendments used for biological control should probably be thoroughly composted before addition to soil in order to avoid stimulation of *Pythium* spp.

Flooding. The destruction of inoculum of soilborne plant pathogens during flooding (Taylor & Guy, 1981; Stover, 1979) is almost certainly accomplished by anaerobes that are active while the propagules are without oxygen and unable to resist decay (Cook & Baker, 1983). The use of paddy rice in a rotation is among the most effective natural treatments ever discovered for control of pathogens in soil. On the other hand, as with the pathogens eradicated by soil fumigation, the flooding may leave a biological vacuum in the soil available for rapid recolonization by surviving or introduced inoculum of pathogens (Stover, 1979). It is rare, therefore, that the benefits of flooding last beyond the first crop after the treatment.

Solarization. Elimination of pathogen inoculum by solarization is a relatively recent innovation (Katan *et al.*, 1976; Katan, 1980). The treatment involves covering the soil surface with clear polyethylene sheeting to trap incoming radiation and thereby elevate the soil temperature. Although much of the pathogen control results from heating the soil above the thermal death point of the organism, part of the effect is also thought to result from increased antagonism from certain soil microorganisms and may be another example of biological control. As with fumigation, this treatment is expensive and is therefore limited to small areas and high value crops. The treatment may also leave a biological vacuum available for recolonization by pathogens.

Introduced Antagonists. While not yet in commercial practice, there is considerable research underway on the use of introduced antagonists for biological destruction of pathogen inoculum. Most of this effect is aimed at pathogens that survive in soil as sclerotia, e.g. *Sclerotinia* spp. and *Sclerotium rolfsii*. The hyperparasites shown to be effective under field conditions include *Sporoidesmium sclerotivorum* (Adams & Ayres, 1982), *Conithyrium minitans* (Tribe, 1957; Turner & Tribe, 1976), and *Trichoderma* spp. (Wells *et al.*, 1972). There is also good progress on the use of *Bacillus penetrans* as a hyperpathogen of nematodes (Stirling & Wachtel, 1980); this antagonist introduced into soil can cause significant reductions in the population of *Meloidogyne* spp. All of these agents are candidates for use as a preplant soil treatment. The main limiting factor in their use is an economical method for mass rearing (Cook & Baker, 1983).

The Use of Microrganisms to prevent infection or limit Disease Development

Some highly effective biological controls are now available or under development that prevent the pathogen from infecting plants or that limit pathogenesis after infection. Some of these involve the use of antagonistic microorganisms that compete with the pathogen in the infection court, inhibit the pathogen by antibiotic action, or parasitize the pathogen mycelium or spores within the infected tissue. Other controls involve avirulent or hypovirulent forms of the pathogen used to compete with the pathogenic forms, induce the host to become more resistant, or vector a mycovirus responsible for a transmissible hypovirulence. These biological controls operate naturally and are subject to effects of cultural practices, but they also provide unlimited opportunities as sources of introduced agents and for application of biotechnology to biological control (Cook, 1985a).

Green plants support a large population of nonpathogenic microorganisms as epiphytes and endophytes on both above- and below-ground organs. The vast majority of these microorganisms exist in a commensal relationship with the plant, deriving their energy supply at virtually no expense to the host. A portion of this superficial microbiota are likely to become the pioneer colonists of the tissue or organ once the plant dies; by their prior establishment on and in the plant while it lives, they are ideally poised to then colonize the plant saprophytically once it dies. However, while on the living plant these nonpathogenic or weakly parasitic microorganisms are also ideally poised to compete with pathogens, occupy infection sites, or colonize the lesions as secondaries and limit reproduction or subsequent survival by the pathogen. As such, they can provide considerable biological protection of the plant that is subject to manipulation.

As a minimum, precautions must be taken not to upset any natural biological protection provided by the nonpathogenic epiphytes and endophytes. Occasionally, a fungicide applied to the foliage for the control of one disease has been observed to favor a previously unimportant disease, because it also eliminated one or more nonpathogens on the phylloplane that had provided a natural biological control (Fokkema et al., 1975; Griffiths, 1981). There is also evidence that different plant species and even cultivars support a specific as well as cosmopolitan nonpathogenic microbiota, suggesting that their presence on plants is under genetic control of the plant and is therefore subject to improvement by plant breeding (Neal et al., 1973; Bird, 1982). However, the most direct and effective means to exploit this protective microbiota is to select out or breed for the most aggressive and inhibitory strains, mass rear them, and then introduce them where needed.

The current research and testing effort on biological control of plant pathogens by microorganisms on and within the plant itself can be grouped under five broad categories:

1) Protection of seeds and seedlings against pre- and post-emergence blight by competitive or antibiotic-producing nonpathogens.
2) Protection of roots and rootlets from damage by root-colonizing micro-organisms.

3) Protection of foliage and blossoms by competitive but nonpathogenic epiphytes.
4) Protection of wounds by nonpathogens.
5) Cross protection or induced resistance with mildly virulent, hypovirulent, or avirulent strains of the pathogen.

Protection of Seeds and Seedlings. A number of nonpathogenic fungi and bacteria have shown efficacy against pathogens of germinating seeds and seedlings when introduced as a biological seed treatment (Kommedahl & Windels, 1981). The pathogens controlled are mainly *Pythium* spp. and *Rhizoctonia solani* responsible for seed decay and damping off, and *Fusarium* spp. responsible for seedling blights. Agents used to protect against these pathogens include: *Trichoderma* spp. (Chet *et al.*, 1981), *Penicillium* spp. (Kommedahl *et al.*, 1981), *Chaetomium globossum* (Kommedahl & Mew, 1975), and *Bacillus subtilus* (Kommedahl & Mew, 1975). At least part of the protective action of these microorganisms results from their ability to use the seed exudates and pre-empt the pathogen of its nutrient source. Some strains of *T. harzianum* also produce enzymes that break down the cell walls of *Pythium* spp., *R. solani* (Chet *et al.*, 1981), or both. Seed and seedling protection by *B. subtilus* and *Penicillium* spp. probably results, in part, from the ability of these antagonists to produce antibiotics inhibitory to the pathogens.

Field trials with biological seed treatments have been conducted successfully with peas, soybeans, corn, and many other crop plants (Kommedahl & Windels, 1978; Kommedahl *et al.*, 1981). Thus far, however, no commercial use has been made of this biological control. A major constraint is lack of interest: chemical seed treatments are highly effective against seed decay and seedling blights and commonly less expensive. Nevertheless, the technology for efficient mass production of agents for their application as seed protectants is becoming available and may be advantageous, especially in countries where treated seed left after sowing is to be eaten or fed to livestock. Biological seed treatments would be safer than chemicals in this regard as most would be inactivated upon cooking.

Protection of Roots. It is relatively easy to find agents that, when applied on the seed, are capable of colonizing the roots for several centimeters into the soil. Particularly noteworthy among these are the root-colonizing fluorescent *Pseudomonas* spp. (Weller, 1985). These bacteria as a group have a high degree of rhizosphere competence, and most produce one or more antibiotics and one or more siderophores. Certain strains of *B. subtilus* and *Streptomyces* spp. are also rhizosphere competent, produce substances inhibitory to pathogens, and have been shown to protect roots against pathogens when applied on seed (Broadbent *et al.*, 1971; Merriman *et al.*, 1975). As with crop rotation, tillage, soil fumigation, and certain organic crop treatments used for root pathogen control, the increased growth response from a biological seed treatment can give the appearance that the plant has been stimulated or that the supply of plant nutrients has been increased, when actually the plant is only growing more nearly at its biological potential because of a healthier more efficient root system.

A major breakthrough in biological seed treatments for root protection was the discovery of Broadbent *et al.* (1971) in Australia of *B. subtilus* strain A 13. This bacterium was one of several hundred isolates obtained initially from various sources, but was unique in being inhibitory (antibiotic) *in vitro* to all nine of nine root pathogens included in a preliminary screening test. This strain as a seed treatment in field trials increased yields of carrots, sweet corn, and cereals and improved seedling vigor and size of several ornamental plants (Merriman *et al.*, 1975). Plant growth "promotion" by seed bacterization had been observed prior to this work (Brown, 1974), but this was probably the first good evidence that ability to produce antibiotics and inhibit root pathogens was an important trait of the bacteria. A similar strain of *B. subtilus* with broad-spectrum antibiotic properties was used as a seed treatment on approximately 5,000 ha of peanuts in Alabama in 1984 (P. Backman, personal communication).

A second breakthrough in use of microorganisms as a seed treatment for root protection was the discovery by Burr *et al.*, (1978), Kloepper *et al.*, (1980b), and Suslow & Schroth, (1982b) of the plant growth promoting rhizobacteria (PGPR). The PGPR are strains of fluorescent *Pseudomonas* spp. having broadspectrum inhibitory activity against root pathogens including by production of siderophores (Kloepper *et al.*, 1980a). The PGPR strains were isolated originally from plant roots (potato, radish), which may account for their rhizosphere competence. Populations of 10^3–$10^5/0.1$ g root tissue were shown to occur on roots of potato by the PGPR applied on potato seed pieces. Moreover, the natural population of fungi and Gram negative bacteria on roots was lower in the presence of PGPR than in their absence (Kloepper & Schroth, 1979). Being aggressive root colonists and able to produce substances inhibitory to other microorganisms, the PGPR either pre-empt or displace the normal rhizosphere microbiota, including exopathogens; the roots are therefore healthier. Yield increases of 5–10% were obtained for potatoes (Kloepper *et al.*, 1980b) and sugar beets (Suslow & Schroth, 1982b). Mutant strains of PGPR unable to produce antibiotic did not result in better plant growth (Kloepper & Schroth, 1981b). This is the first direct experimental evidence that ability to inhibit pathogens by production of antibiotic is an important trait of the agents used as seed treatments for root protection.

Upon the discovery 40–50 years ago that antibiotic-producing micro-organisms occurred in abundance in soil, plant pathologists attempted to isolate, mass-rear, and reintroduce them into soil for biological control of root pathogens. The experiments failed so completely that virtually all such research was abandoned until recently. The difference in approach now compared with years ago is relatively simple. Earlier workers failed to take into account that microorganisms occupy ecological niches in soil and cannot readily be introduced as aliens into niches where they are not adapted. The adjustment has been to isolate the antibiotic-producing types from the rhizosphere rather than bulk soil and then reintroduce them on the planting material so they can re-establish in the rhizosphere.

Weller & Cook (1983) targeted their strains of root-colonizing psuedomonads for biocontrol of *Gaeumannomyces graminis* var. *triciti* and *Pythium* spp. responsible, respectively, for take-all and pythium root rot. They obtained their strains from the healthiest roots available in soil infested with one or the other of the targeted pathogens and then screened *in vitro* for ability of the strains to produce antibiotics effective against one or both pathogens. In the case of take-all, the antagonists were isolated from roots of wheat growing in soil from fields that had undergone take-all decline. The evidence of Cook & Rovira (1976) and Weller (*in* Cook & Baker, 1973) suggests that take-all decline results from a natural build-up and eventual protection of roots by inhibitory pseudomonads. These bacteria establish not only on roots, but more particularly in lesions, where they have the potential to limit *G. graminis* var. *tritici* not only during its parasitic existence but also as a cohabitant with the pathogen in infested crop residue. The isolation of rhizobacteria from roots naturally protected by rhizobacteria increased the chances of finding effective strains. Their procedure also helps insure that the biocontrol agents will be adapted not just to wheat roots but to wheat roots inoculated or even infected with the target pathogen, the situation most likely to be encountered in the field. By their approach, strains of fluorescent *Pseudomonas* spp. were obtained that, when applied on seeds, gave 10–25% increased yield in commercial fields where take-all was the yield-limiting factor (Weller & Cook, 1983). One strain inhibitory to *Pythium* spp. increased the yield of wheat in field plots at Pullman, Washington, by 27% where pythium root rot was the yield-limiting factor (Weller & Graham, 1984).

While some of these microorganisms may well produce growth factors or make nutrients available to the plant, it remains to be shown that mutant strains unable to produce these factors are also unable to promote plant growth. In contrast, strains unable to produce antibiotics and/or siderophores are unable to promote plant growth (Becker & Cook, 1984; Kloepper & Schroth, 1981b), and the strains have no effect on plants where soil fumigation is used to eliminate pathogens (Kloepper & Schroth, 1981a; Weller & Cook, 1983). Geels & Schippers (1983) showed further that root-colonizing pseudomonads had no beneficial effect on potatoes in plots where potatoes were grown only once every three years, but they resulted in greater yield of marketable tubers where potatoes were grown in monoculture and yields had declined. This observation makes clear the importance of testing where the target root pathogen is the yield-limiting factor, and the futility of expecting a growth response from a crop by treatment with rhizobacteria if the root pathogens are already controlled by some other method, e.g. crop rotation or soil fumigation. All of these data and observations verify the importance of root disease control to crop productivity but suggest further that biological seed treatments have the potential to protect roots in a way that presently can be achieved by soil fumigation or crop rotation.

Plant Protection by Agents on the Foliage. Protection of above-ground plant parts by microorganisms sprayed on the foliage likewise is not yet a commercial reality but could be in the near future. As with seed and root protection, the introduced agent must be able to occupy and preferably dominate the ecological niche of the pathogen. The pathogens most vulnerable to control are those that depend on an exogenous nutrient supply from the plant for spore germination and/or prepenetration growth. An aggressive leaf colonist, nonpathogenic itself but able to use the leaf exudates or pollen needed by the pathogen, can pre-empt such pathogens and provide significant biological control (Blakeman & Fokkema, 1982). An example is the biological control of ice-nucleation active (INA) strains of *Pseudomonas syringae* by prior establishment on the leaves of INA-negative strains of either *P. syringae* or *Erwina herbicola* (Lindow, 1983). Strains of INA-negative *P. syringae* have now been developed by genetic engineering, where only the *ice* gene was removed from an INA-positive strain. Apple and pear blossoms can be partially protected against the fireblight bacterium, *Erwina amylovora*, by introduction of either the nonpathogenic *E. herbicola* or *P. fluorescens* (Beer *et al.*, 1980), and the leaf scars on apple trees can be protected against infection by *Nectria galligena* by prior application of *Bacillus subtilus* (Swinburn, 1973). The bacteria used to control fireblight are thought to pre-empt the fireblight pathogen and thereby limit its nutrient supply. *B. subtilus* probably limits *N. galligena* by antibiotic action (Swinburn *et al.*, 1975).

Protection of Wounds. The two best-known examples of biological control of pathogens by an introduced microorganism involve protection of wounds against infections. One is the use of *Peniophora gigantea* applied to freshly cut stumps of pine to protect the cut surface from infection by airborne spores of *Heterobasidion annosum* (Rishbeth, 1963; 1979). Without protection, *H. annosum* can colonize the entire stump and then use it as a foodbase to infect nearby trees. *P. gigantea* as a weak parasite establishes only superfically on the cut surface, but this is adequate to prevent infection by *H. annosum*. The mechanism of antagonism may involve hyphal interference (Ikediugwu *et al.*, 1970). The other example is the use of the avirulent *Agrobacterium radiobacter* strain K84 applied as a dip to bare roots of fruit-tree and ornamental transplants to protect the roots against crown gall caused by *A. radiobacter* pv. *tumefaciens* (Kerr, 1980). The crown gall pathogen is soilborne and infects through wounds caused during the transplanting. Strain K84 established in the wounds prior to transplanting in the orchard or garden prevents infections by the virulent strain. The protection results from occupancy of infection sites in the wounds by the avirulent strain, and also from production of a bacteriocin by the avirulent strain to which the virulent strain is sensitive. Both of these methods of biocontrol are in commercial use in several countries (Moore, 1979).

Pruning wounds serve as portals of entry for plant pathogens and can be protected by prior establishment of a suitable antagonist on the wound surface. An example is the use in Australia of the weak pathogen *Fusarium lateritium* applied on fresh pruning wounds on apricot to protect against airborne inoculum of *Eytypa armenaceae*, the gumosis pathogen (Carter & Price, 1974). Colonization of the wound surface by the *Fusarium* is sometimes too slow to provide immediate protection, but this problem can be solved by applying a mixture of *Fusarium* spores and benomyl (Carter, 1983); the fungicide provides immediate protection and the benomyl-tolerant *Fusarium* provides long-term protection. Another example is the use in England and elsewhere in Europe of *Trichoderma* spores to protect pruning wounds on plum trees against infection by *Stereum purpureum*, cause of silver leaf (Corke, 1974). The introduction of *Trichoderma* as colonized wood dowels driven into infected trees may even cause remission of the silver leaf disease—a possible biotherapeutic effect.

A particularly novel biocontrol whereby an active disease goes into remission following introduction of the biocontrol agent is the control in France of chestnut blight by hypovirulent strains of the pathogen, *Endothia parasitica* (Grente & Sauret, 1980a & 1980b). Hypovirulence in *E. parasitica* is transmissable and is associated with one or more dsRNA mycovirus-like agents carried in the fungus (Day *et al.,* 1977). When a hypovirulent strain is introduced into an active canker and is vegetatively compatible (hyphae anastomose) with the virulent pathogen in the canker, the pathogen then becomes hypovirulent and the canker heals. Hypovirulence is spreading naturally in chestnut stands in southern France and northern Italy, and the introduction of appropriate strains into the remaining active cankers is expected to bring the disease under control more quickly (Anasnostakis, 1982).

Enhanced or Induced Resistance and Cross Protection. This approach to plant protection with microorganisms involves the prior establishment of a non-pathogen, weak strain of a pathogen, or a mycorrhizal fungus in the plant to protect against a virulent pathogen. Ectomycorrhizae on pine roots can provide protection of pines against *Phytophthora cinnomomi* (Marx, 1975) and endomycorrhizae can enhance the resistance of some plants to certain root pathogens (Schönbeck & Dehne, 1979), especially if established in the root well in advance of a challenge inoculation (Bärtschi *et al.*, 1981). Avirulent or weak strains of the pathogen, on the other hand, may induce resistance (Kuć, 1982) or provide cross protection (Hamilton, 1980) against more virulent strains of the pathogen. Induced resistance and cross protection are about the only way microorganisms can be used to protect plants against certain vascular pathogens and systemic viruses. Cross protection is a general term for a phenomenon whereby the first agent introduced into the plant tissues confers protection of the tissues against a second agent. Induced resistance is more explicit: the first agent triggers a resistance response in the plant effective against the second agent.

Cross protection has been used for several years in Europe, Japan, and the People's Republic of China to control tobacco mosaic virus in tomato (Fletcher & Butler, 1975; Fletcher & Rowe, 1975). It has also been used in Brazil and Australia to control citrus triteza virus in orange trees (Costa & Muller, 1980). In both cases healthy, virus-free seedlings are inoculated with a mild strain of the virus, which then protects the plant for several years against more severe strains of the virus. Trees may be less vigorous and yield less because of the mild strain, but they would be even less vigorous and yield even less if infected by a virulent strain of the virus in the absence of any protection. More work is needed on the mechanisms of cross protection and on finding better but less damaging strains so as to improve the safety and reliability of this approach to biological control. Plant viruses are presently responsible for widespread crop damage and many spread rapidly, either by insect vectors or in plant sap on tools and hands of workers. Resistant cultivars often are not available and methods used to prevent virus spread are either ineffective or impractical. Cross protection or induced resistance with mild strains is as practical as any known method of control and could become a common and highly effective procedure if the risks could be lowered.

Pathogens that cause vascular wilt, e.g. *Fusarium oxysporum* and *Verticillium* spp., sometimes can be controlled by prior inoculation of the plant with a nonpathogenic strain related to the pathogen. Usually the control is only temporary (Wymore & Baker, 1982), but Ogawa & Kommada (1984) in Japan showed that sweet potato inoculated as cuttings with a nonpathogenic strain of *Fusarium oxysporum* were protected until harvest against *F. oxysporum* f. sp. *batatas*. The biocontrol strain was isolated originally from the xylem of a healthy sweet potato plant growing where fusarium wilt was unexpectedly mild. Stems of sweet potato were systematically protected by the nonpathogenic strain, suggesting that induced resistance was somehow involved. *Fusarium oxysporum* can be mass produced easily as bud cells in liquid culture, and the normal procedure for sweet potatoes in Japan is to transplant cuttings. The prospects for commercial production and use of *F. oxysporum* to protect sweet potatoes in Japan are thus highly promising.

In many wheat-growing areas of the world, *G. graminis* var. *tritici* occurs at a high inoculum potential on roots of grass, and wheat following the grass may, therefore, exhibit severe take-all the first year (Sprague, 1950). However, in England, *Phialophora graminicola* (sexual stage=*Gaeumannomyces cylin-*

drosporum) present on roots of grass in a ley, provides significant natural protection against take-all of the first one or two wheat crops after the grass ley (Deacon, 1976). *P. graminicola* is avirulent or only weakly virulent on roots of wheat under conditions in England, but it nevertheless colonizes the cortex of wheat roots and subsequently protects them against the take-all pathogen (Speakman & Lewis, 1978). As with other uses of induced resistance and cross protection, the use of *P. graminicola* for take-all control may not be without risk. *P. graminicola* is pathogenic on wheat when soil temperatures exceed 30°C (R.W. Smiley, personal communication). The use of *P. graminicola* for biocontrol of take-all may therefore be limited in use to cooler wheat-growing areas.

BIOLOGICAL, CONTROL, A BROAD CONCEPT

Biological control has been broadly defined (Cook & Baker, 1983) to include pathogen control not only by microorganisms but also by higher plants and even by the pathogen used against itself. This broad definition is an obvious departure from that used in entomology—the use of pathogens, parasites, and predators to regulate the population of an insect pest (DeBach, 1964). However, a broader concept is needed in plant pathology where the opportunities for biological control include not only regulation of the pathogen population but also restriction of the disease-producing activities (pathogensis) of a pathogen on and within the host. The biological control of *Heterobasidion annosum* on pine by *Peniophora gigantea* applied to freshly cut stumps, and of crown gall by strain K84 of *Agrobacterium radiobacter* applied to bare roots of transplants, both are examples of an antagonist applied to protect the host against infection; whether or not the pathogen population is diminished is of little or no consequence. Moreover, the host provides the "battlefield" for pathogen-antagonist interaction and as such, plays a passive role in the control. However, there is no logical reason to limit the concept of biological control to examples of a passive role of the host but exclude examples of an active role of the host, i.e. active host plant resistance (Cook, 1985a). It is also illogical to include resistance induced by a microorganism as an example of biological control but exclude resistance achieved by genetic modification of the host; or to include enhanced resistance by mycorrhizae as an example of biocontrol but exclude enhanced resistance achieved with a cultural practice. Lupton (1983) stated in his presidential address to the Association of Applied Biologists: "accelerating or diverting evolutionary processes in order to obtain genotypes adapted to [man's] needs are a most important example of the application of biological control to agricultural and horticultural crops."

Perhaps the best reason for including active host resistance to pathogens as an example of biological control is that, to do otherwise would be counterproductive in the face of developments anticipated for the future (Cook, 1985a). Consider, for example, a gene that codes for the synthesis of a substance inhibitory to a plant pathogen. Our concept of biocontrol must not be so confining that the use of this gene in a microorganism to control a pathogen on the host can be labeled biocontrol, but its use in the plant to inhibit the same pathogen cannot be labeled as biocontrol. The biocontrol of the future will be based on molecular biology as well as ecology and microbiology, and will involve genetic improvements of antagonists, plants, or both to produce the least favorable if not most inhibitory plant habitat for pathogens. The biocontrol of the future will also see greater use of the pathogen, genetically modified and introduced to compete with more virulent forms, induce host plant resistance, vector a hypovirulence agent, or produce a specific inhibitor. Genes for avirulence or self inhibition in the pathogen and antibiotic production by antagonists may also serve someday as genetic sources of resistance in plants. It is this broad concept of biological control that makes this approach so fascinating and unconfining as a field of study and so promising as an approach to disease control compatible with a sustainable agriculture.

SUMMARY

The means to protect plants from diseases and thereby permit them to grow and produce near or at their potential will be increasingly more important in the future. It is not compatible with a sustainable agriculture to allow diseases to limit crop yields so significantly relative to the potential yields as set by the weather, climate, and agronomic inputs. For complete control of diseases of the growing plant, the planting material (seeds or transplants) must be pathogen free, planted in pathogen-free soil, and the foliage must be protected for the entire vegetative period. The severe yield-limiting effect of soilborne pathogens on crops is readily demonstrated by the often spectacular response of the crop to soil fumigation; it resembles a response to added fertilizer but is the result of a healthier root system more efficient in uptake of nutrients. Yields invariably decline with practices that favor root disease, e.g. crop monoculture, and they are higher in response to practices that eliminate or suppress pathogen inoculum in soil, e.g. fumigation or heating the soil by solarization or burning a layer of crop residue. Crop rotation, tillage, temporary flooding of the soil, certain organic amendments, and possibly solarization are practices that exploit the natural biological control of pathogen inoculum by the microorganisms in soil. Some progress has been made in the use of hyperparasites and hyperpathogens introduced into soil to reduce the inoculum density of plant pathogens. Beneficial microorganisms can also be used to protect plants from infection or limit disease progress after infection. Considerable protection is afforded by the natural populations of nonpathogenic epiphytes and endophytes on plants and progress has been made on techniques to establish the more effective species or strains on or in plant parts where needed, e.g. by introduction on seeds or application to the foliage or in wounds caused by pruning or transplanting. In some cases plants may be inoculated with a mild or avirulent strain of the pathogen to cross protect against a more virulent strain of the pathogen. Biological control of plant pathogens is a broad concept and includes the use of host resistance as well as beneficial microorganisms, manipulated genetically or by adjustments in the environment.

References

Adams, P.B. & Ayres, W.A. (1982). Biological control of Sclerotinia lettuce drop in the field by *Sporidesmium sclerotivorum. Phytopathology,* **72**, 485–588.

Anagnostakis, S.L. (1982). Biological control of chestnut blight. *Science,* **215**, 466–471.

Baker, K.F. & Cook, R.J. (1974). Original edn. *Biological Control of Plant Pathogens.* W.H. Freeman; San Francisco. Reprinted edn.(1982). American Phytopathological Society; St. Paul, MN. 433 pp.

Baker, K.F. & Snyder, W.C. eds. (1965). *Ecology of Soil-Borne Plant Pathogens. Prelude to Biological Control.* University of California Press; Berkeley. 571 pp.

Bärtschi, H., Gianinazzi-Pearson, V. & Vegh, I. (1981). Vesicular-arbuscular mycorrhiza formation and root rot disease (*Phytophthora cinnamomi*) development in *Chamaecyparis lawsoniana. Phytopathologische Zeitschrift,* **102**, 213–218.

Becker, O.J. & Cook, R.J. (1984). *Pythium* control by siderophore-producing bacteria on roots of wheat. *Phytopathology,* **74**, 806.

Beer, S.V., Norelli, J.L., Rundle, J.R., Hodges, S.S., Palmer, J.R., Stein, J.I. & Aldwinkle, H.S. (1980). Control of fire blight by non-pathogenic bacteria. *Phytopathology,* **70**, 459.

Bird, L.S. (1982). The MAR (multi-adversity resistance) system for genetic improvement of cotton. *Plant Disease,* **66**, 172–176.

Blakeman, J.P. & Fokkema, N.J. (1982). Potential for biological control of plant disease on the phylloplane. *Annual Review of Phytopathology,* **20**, 167–192.

Broadbent, P. & Baker, K.F. (1975). Soils suppressive to Phytophthora root rot in eastern Australia. In *Biology and Control of Soil-Borne Plant Pathogens* (G.W. Bruehl, ed.), pp. 152–157. American Phytopathological Society; St. Paul, MN. 216 pp.

Broadbent, P., Baker, K.F. & Waterworth, Y. (1971). Bacteria and actinomycetes antagonistic to fungal root pathogens in Australian soils. *Australian Journal of Biological Sciences,* **24**, 925–944.

Brown, M.E. (1974). Seed and root bacterization. *Annual Review of Phytopathology,* **12**, 181–197.

Browning, J.A. (1983). Goal for plant health in the age of plants: A national plant health system. In *Challenging Problems in Plant Health.* (T. Kommedahl & P.H. Williams, eds.), pp. 45–57. American Phytopathological Society; St. Paul, MN.

Bruehl, G.W., ed. (1975). *Biology and Control of Soil-borne Plant Pathogens.* American Phytopathological Society; St. Paul, MN. 216 pp.

Burr, T.J., Schroth, M.N. & Suslow, T. (1978). Increased potato yields by treatment of seedpieces with specific strains of *Pseudomonas fluorescens* and *P. putida. Phytopathology,* **68**, 1377–1383.

Carter, M.V. (1983). Biological control of *Eutypa armeniacae.* V. Guidelines for establishing routine wound protection in commercial apricot orchards. *Australian Journal of Experimental Agriculture and Animal Husbandry,* **23**, 429–436.

Carter, M.V. & Price, T.V. (1974). Biological control of *Eutypa armeniacae*. II. Studies of the interaction between *E. armeniacae* and *Fusarium lateritium* and their relative sensitivities to benzimidazole chemicals. *Australian Journal of Agricultural Research*, **25**, 105–109.

Chamswarng, C. & Cook, R.J. (1985). Identification and comparative pathogenicity of *Pythium* species from wheat roots and wheat-field soils in the Pacific Northwest. *Phytopathology*, **75**, 821–827.

Chet, I. & Baker, R. (1981). Isolation and biocontrol potential of *Trichoderma hamatum* from soil naturally suppressive to *Rhizoctonia solani*. *Phytopathology*, **71**, 286–290.

Chet, I., Harman, G.E. & Baker, R. (1981). *Trichoderma hamatum*: Its hyphal interaction with *Rhizoctonia solani* and *Pythium* spp. *Microbial Ecology*, **7**, 29–38.

Cook, R.J. (1984a). Root health: Importance and relationship to farming practices. In *Organic Farming: Current Technology and its Role in a Sustainable Agriculture* (D.F. Bezdicek, J.F. Power, D.R. Kenney & M.J. Wright, eds.), pp. 111–127. American Society of Agronomy Special Publication No. 46; Madison, WI. 192 pp.

Cook, R.J. (1984b). Biological control of root pathogens: New technologies and the potential for impact on crop productivity. In *Soilborne Crop Diseases in Asia*, pp. 206–214. Food and Fertilizer Technology Center for the Asian and Pacific Region Book Series No. 26. 240 pp.

Cook, R.J. (1985a). Biological control of plant pathogens: Theory to application. *Phytopathology*, **75**, 25–29.

Cook, R.J. (1985b). Pathogens as constraints to crop productivity. In *Water and Water Quality in World Food Policy* (W.R. Jordan, ed.) Texas A & M University; College Station, TX. (In press).

Cook, R.J. & Baker, K.F. (1983). *The Nature and Practice of Biological Control of Plant Pathogens*. American Phytopathological Society; St. Paul, MN. 439 pp.

Cook, R.J. & W.A. Haglund. (1982). Pythium root rot: A barrier to yield of Pacific Northwest wheat. Washington State University Agricultural Research Center Bulletin No. XB0913. 20 pp.

Cook, R.J. & Rovira, A.D. (1976). The role of bacteria in the biological control of *Gaeumannomyces graminis* by suppressive soils. *Soil Biology & Biochemistry*, **8**, 267–273.

Cook, R.J., Boosalis, M.G. & Doupnik, B. (1978). Influence of crop residues on plant diseases. In *Crop Residue Management Systems* (W.R. Oschwald, ed.), pp. 147–163. American Society of Agronomy Special Publication No. 31; Madison, WI. 248 pp.

Cook, R.J., Sitton, J.W. & Waldher, J.T. (1980). Evidence for *Pythium* as a pathogen of direct-drilled wheat in the Pacific Northwest. *Plant Disease*, **64**, 102–103.

Corke, A.T.K. (1974). The prospects for biotherapy in trees infected by silver leaf. *Journal of Horticultural Science*, **49**, 391–394.

Costa, A.S. & Muller, G.W. (1980). Tristeza control by cross protection: A U.S.-Brazil cooperative success. *Plant Disease*, **64**, 538–541.

Day, P.R., Dodds, J.A., Elliston, J.E., Raynes, R.A. & Anagnostakis, S.L. (1977). Double-stranded RNA in *Endothia parasitica*. *Phytopathology*, **67**, 1393–1396.

Deacon, J.W. (1976). Biological control of the take-all fungus, *Gaeumannomyces graminis* by *Phialophora radicicola* and similar fungi. *Soil Biology & Biochemistry*, **8**; 275–283.

DeBach, P., ed. (1964). *Biological Control of Insect Pests and Weeds*. Reinhold; New York. 844 pp.

Elliott, L.F., McCalla, M.T. & Waiss, A. Jr. (1978). Phytotoxicity associated with residue management. In *Crop Residue Management Systems* (W.R. Oschwald, ed.), pp. 131–146. American Society of Agronomy Special Publication No. 31; Madison, WI. 248 pp.

Fletcher, J.T. & Butler, D. (1975). Strain changes in populations of tobacco mosaic virus from tomato crops. *Annals of Applied Biology*, **81**, 409–412.

Fletcher, J.T. & Rowe, J.M. (1975). Observations and experiments on the use of an avirulent mutant strain of tobacco mosaic virus as a means of controlling tomato mosaic. *Annals of Applied Biology*, **81**, 171–179.

Fokkema, N.J., van de Laar, J.A.J., Nelis-Blomberg, A.L. & Schippers, B. (1975). The buffering capacity of the natural mycoflora of rye leaves to infection by *Cochliobolus sativus*, and its susceptibility to benomyl. *Netherlands Journal of Plant Pathology*, **81**, 176–186.

Garrett, S.D. (1956). *Biology of Root-Infecting Fungi*. Cambridge University Press; London. 294 pp.

Geels, F.P. & Schippers, B. (1983). Reduction of yield depression in high frequency potato cropping after seed tuber treatments with antagonistic fluorescent *Pseudomonas* spp.

Phytopathologische Zeitschrift, **108**, 207–214.

Grente, J. & Sauret, S. (1969a). L'hypovirulence exclusive phénomène original en pathologie végétale. *Compte rendu hebdomadaire des seances de l'Academie des sciences*, **268**, 2347–2350.

Grente, J. & Sauret, S. (1969b). L'hypovirulence exclusive est-elle contrôlée par des déterminants cytoplasmiques? *Compte rendu hebdomadaire des seances de l'Academie des sciences*, **268**, 3173–3176.

Griffiths, E. (1981). Iatrogenic plant diseases. *Annual Review of Phytopathology*, **19**, 69–82.

Hamilton, R.H. (1980). Defenses triggered by previous invaders-viruses. In *Plant Disease. An Advanced Treatise* (J.G. Horsfall & E.B. Cowling, eds.), Vol. V, Chapter 12. Academic Press; New York.

Hwang, S.C., Chen, C.L. & Lin, J.C. (1984). Cultivation of banana using plantlets from meristem culture. *HortScience*, **19**, 231–233.

Ikediugwu, F.E.O., Dennis, C. & Webster, J. (1970). Hyphal interference by *Peniophora gigantea* against *Heterobasidion annosum*. *Transactions of the British Mycological Society*, **54**, 307–309.

Kaiser, W.J. & Horner, G.M. (1980). Root rot of irrigated lentils in Iran. *Canadian Journal of Botany*, **58**, 2549–2566.

Katan, J. (1980). Solar pasteurization of soils for disease control: Status and prospects. *Plant Disease*, **64**, 450–454.

Katan, J., Greenberger, A., Alon, H. & Grinstien, A. (1976). Solar heating by polyethylene mulching for the control of diseases caused by soil-borne pathogens. *Phytopathology*, **66**, 683–688.

Kelman, A. & Cook, R.J. (1977). Plant pathology in the People's Republic of China. *Annual Review of Phytopathology*, **15**, 409–429.

Kerr, A. (1980). Biological control of crown gall through production of agrocin 84. *Plant Disease*, **64**, 25–30.

King, C.J., Hope, C. & Eaton, E.D. (1934). Some microbiological activities affected in manurial control of cotton root rot. *Journal of Agricultural Research*, **49**, 1093–1107.

Kloepper, J.W. & Schroth, M.N. (1979). Plant growth promoting rhizobacteria: Evidence that the mode of action involves root microflora interactions. *Phytopathology*, **69**, 1034.

Kloepper, J.W. & Schroth, M.N. (1981a). Plant growth-promoting rhizobacteria and plant growth under gnotobotic conditions. *Phytopathology*, **71**, 642–644.

Kloepper, J.W. & Schroth, M.N. (1981b). Relationship of *in vitro* antibiosis of plant growth-promoting rhizobacteria to plant growth and the displacement of root microflora. *Phytopathology*, **71**, 1020–1024.

Kloepper, J.W., Leong, J., Teintze, M. & Schroth, M.N. (1980a). Enhanced plant growth by siderophores produced by plant growth-promoting rhizobacteria. *Nature*, **286**, 885–886.

Kloepper, J.W., Schroth, M.N. & Miller, T.D. (1980b). Effects of rhizosphere colonization by plant growth-promoting rhizobacteria on potato plant development and yield. *Phytopathology*, **70**, 1078–1082.

Kommedahl, T. & Mew, I.C. (1975). Biocontrol of corn root infection in the field by seed treatment with antagonists. *Phytopathology*, **65**, 296–300.

Kommedahl, T. & Windels, C.E. (1978). Evaluation of biological seed treatment for controlling root diseases of pea. *Phytopathology*, **68**, 1087–1095

Kommedahl, T. & Windels, C.E. (1981). Introduction of microbial antagonists to specific courts of infection: seeds, seedlings, wounds. In *Biological Control of Crop Production* (G. Papavizas, ed.), pp. 227–248. Allanheld, Osmun, Granada; London.

Kommedahl, T., Windels, C.E., Sargini, G. & Wiley, H.B. (1981). Variability in performance of biological and fungicidal seed treatments in corn, peas, and soybeans. *Protection Ecology*, **3**, 55–61.

Kuć, J. (1982). Induced immunity to plant disease. *Bioscience*, **32**, 854–860.

Latin, R.X., Harder, R.W. & Weise, M.V. (1982). Incidence of Cephalosporium stripe as influenced by winter wheat management practices. *Plant Disease*, **66**, 229–230.

Lindow, S.E. (1983). Methods of preventing frost injury caused by epiphytic ice nucleation active bacteria. *Plant Disease*, **67**, 327–333.

Lupton, F.G.H. (1984). Biological control: the plant breeder's objective. *Annals of Applied Biology*, **104**, 1–16.

Marx, D.H. (1975). The role of ectomycorrhizae in the protection of pine from root infection by

Phytophthora cinnamomi. In *Biology and Control of Soil-Borne Plant Pathogens* (G.W. Bruehl, ed.), pp. 112–115. American Phytopathological Society; St. Paul, MN.

Merriman, P.R., Price, R.D., Baker, K.F., Kollmorgen, J.F., Piggott, T. & Ridge, E.H. (1975). Effect of *Bacillus* and *Streptomyces* spp. applied to seed. In *Biology and Control of Soil-Borne Plant Pathogens* (G.W. Bruehl, ed.), pp. 130–133. American Phytopathological Society; St. Paul, MN.

Millard, W.A. & Taylor, C.B. (1927). Antagonism of micro-organisms as the controlling factor in the inhibition of scab by green-manuring. *Annals of Applied Biology,* **14**, 202–216.

Moore, K.J. & Cook, R.J. (1984). Direct-drilling increases take-all of wheat in the Pacific Northwest. *Phytopathology,* **74**, 1044–1049.

Moore, L.W. (1979). Practical uses and success of *Agrobacterium radiobacter* strain 84 for crown gall control. In *Soil-Borne Plant Pathogens* (B. Schippers & W. Gams, eds.), pp. 553–661. Academic Press; New York.

Neal, J.L., Jr., Larson, R.I. & Atkinsin, T.G. (1973). Changes in rhizosphere populations of selected physiological groups of bacteria related to substitution of specific pairs of chromosomes in spring wheat. *Plant & Soil,* **39**, 209–212.

Ogawa, K. & Kommada, H. (1984). Biological control of Fusarium wilt of sweet potato by non-pathogenic *Fusarium oxysporum*. *Annals of the Phytopathology Society of Japan,* **50**, 1–8.

Papavizas, G.C., ed. (1981). *Biological Control in Crop Production.* Beltsville Symposium in Agricultural Research 5. Allanheld, Osmun Co.; London. 461 pp.

Papavizas, G.C. & Lumsden, R.D. (1980). Biological control of soilborne fungal propagules. *Annual Review of Phytopathology,* **18**, 839–413.

Power, J.F. & Legg, J.O. (1978). Effect of crop residues on the soil chemical environment and nutrient availability. In *Crop Residue Management Systems* (W.R. Oschwald, ed.), pp. 85–100. American Society of Agronomy Special Publication No. 31; Madison, WI.

Rishbeth, J. (1963). Stump protection against *Fomes annosus*. III. Inoculation with *Peniophora gigantea*. *Annals of Applied Biology,* **52**, 63–77.

Rishbeth, J. (1979). Modern aspects of biological control of *Fomes* and *Armillaria*. *European Journal of Forest Pathology,* **9**, 331–340.

Sanford, G.B. (1926). Some factors affecting the pathogenicity of *Actinomyces scabies*. *Phytopathology,* **16**, 525–547.

Schippers, B. & Gams, W. eds. (1979). *Soil-Borne Plant Pathogens.* Academic Press; New York. 686 pp.

Schönbeck, F. & Dehne, H.W. (1979). Untersuchungen zum Einfluss der endotrophen Mykorrhiza auf Pflanzenkrankheiten. *Zeitschrift fur Pflanzenkrankheiten, Pflanzenpathologie und Pflanzenschutz,* **86**, 103–112.

Slack, S.A. (1980). Pathogen-free plants by meristem culture. *Plant Disease,* **64**; 15–17.

Speakman, J.B. & Lewis, B.G. (1978). Limitation of *Gaeumannomyces graminis* by wheat root responses to *Phialophora radicicola*. *New Phytologist,* **80**, 373–380.

Sprague, R. (1950). *Diseases of Cereals and Grasses in North America.* Ronald Press; New York. 538 pp.

Stirling, G.R. & Wachtel, M.F. (1980). Mass production of *Bacillus penetrans* for the biological control of root-knot nematodes. *Nematologica,* **26**, 308–312.

Stover, R.H. (1979). Flooding of soil for disease control. In *Soil Disinfestation* (D. Mulder, ed.), pp. 19–28. Elsevier Scientific Publishing; Amsterdam.

Sun, S.K. & Huang, J.W. (1985). Formulated soil amendment for controlling Fusarium wilt and other soilborne diseases. *Plant Disease,* **69**, 917–920.

Suslow, T.V. & Schroth, M.W. (1982a). Role of deleterious rhizobacteria as minor pathogens in reducing crop growth. *Phytopathology,* **72**, 111–115.

Suslow, R.V. & Schroth, M.W. (1982b). Rhizobacteria of sugar beets: Effects of seed application and root colonization on yield. *Phytopathology,* **72**; 199–206.

Swinburne, T.R. (1973). Microflora of apple leaf scars in relation to infection by *Nectria galligena*. *Transactions of the British Mycological Society,* **60**, 389–403.

Taylor, J.B., & Guy, E.M. (1981). Biological control of root-infecting basidiomycetes by species of *Bacillus* and *Clostridium*. *New Phytologist,* **87**, 729–732.

Toussoun, R.A., Bega, R.V. & Nelson, P.E. eds. (1970). *Root Diseases and Soil-Borne Pathogens.* University of California Press, Berkeley. 252 pp.

Tribe, H.T. (1957). On the parasitism of *Sclerotina trifoliorum* by *Coniothyrium mínitans.* *Transactions of the British Mycological Society,* **40**, 489–499.

Turner, G.J. & Tribe, H.T. (1976). On the *Coniothyrium minitans* and its parasitism of *Sclerotinia* species. *Transactions of the British Mycological Society,* **66**, 97–105.

Weller, D.M. (1985). Application of fluorescent pseudomonads to control root diseases. In *Ecology and Management of Soil-Borne Plant Pathogens* (C.A. Parker, K.J. Moore, P.T.W. Wong, A.D. Rovira & J.F. Kollmorgen, eds.), pp. 137–140. American Phytopathological Society; St. Paul, MN.

Weller, D.M. & Cook, R.J. (1983). Suppression of take-all of wheat by seed treatment with fluorescent pseudomonads. *Phytopathology,* **73**, 463–469.

Weller, D.M., Cook, R.J., MacNish, E.W., Powelson, R.L. & Peterson, R.R. (1985). Rhizoctonia bare patch of small grains favored by reduced tillage in the PNW. *Plant Disease,* **69**, 000–000.

Weller, D.M. & Graham, M.C. (1984). Application of fluorescent pseudomonads to improve the growth of wheat. *Phytopathology,* **74**, 806.

Wells, H.D., Bell, D.K. & Jaworski, C.A. (1972). Efficacy of *Trichoderma harzianum* as a biocontrol for *Sclerotium rolfsii. Phytopathology,* **62**, 442–447.

Wilhelm, S. (1965). *Pythium ultimum* and the soil fumigation growth response. *Phytopathology,* **55**, 1016–1020.

Wilhelm, S. & Paulus, A.O. (1980). How soil fumigation benefits the California strawberry industry. *Plant Disease,* **64**, 264–270.

Williams, P.H. (1979). Vegetable crop protection in the People's Republic of China. *Annual Review of Phytopathology,* **17**, 311–324.

Wymore, L.A. & Baker, R. (1982). Factors affecting cross-protection in control of Fusarium wilt of tomato. *Plant Disease,* **66**, 908–910.

Zadocks, J.D. & Schein, R.D. (1979). *Epidemiology and Plant Disease Management.* Oxford University Press; New York. 427 pp.

The Contributions of Fungi, Bacteria and Physical Processes in the Development of Aggregate Stability of a Cultivated Soil

M.B. Molope[1] and E.R. Page[2]

[1]Department of Environmental Science and [2]Department of Biological Science, University of Stirling, Stirling, FK9 4LA, Scotland

INTRODUCTION

The primary mineral particles of soil (clay$<2\mu$m equivalent diameter, sand$>63\,\mu$m, and silt particles between these limits) determine soil texture. These primary particles are bound together into aggregates, thus giving structure to the soil. The stability of aggregates is conveniently measured by their ability to withstand wetting by water under defined conditions.

Aggregate stability of soil improves under pasture, or after addition of organic matter. This structural improvement is mainly brought about by the activities of soil organisms in transforming root debris or organic amendments. Metabolic products are well placed for adsorption onto or reaction with clay colloids. Some organic products, polysaccharides, for example, are short-lived, others, like humic substances, are long-lasting. When soil is cultivated some of the macromolecules may be degraded by oxidation processes, and the aggregates become less stable. On the other hand, increased microbial activity may produce fresh adhesive substances. In addition certain physical processes also affect soil aggregate stability.

We have studied these dynamic processes in several English and Scottish soils. The typical results presented here refer only to one of the soils, an example from the Bromyard series.

MATERIALS AND METHODS

The Bromyard series soils are red brown earths, with good agricultural

TABLE 1

Characteristics of Bromyard series soils

Samples	Clay <2μm	Silt 2-63 μm	Sand 63μm-2mm	Organic carbon	pH	Plastic limit	Cultivation history
Bromyard 1	18.2	65.8	9.1	3.7	6.8	31.7	Permanent pasture (more than 50 years)
Bromyard 2	13.4	69.5	10.5	2.0	7.1	25.5	Continuous arable

potential, but continuous cultivation depletes the organic matter, and the consequent loss of stability gives rise to structural defects which can limit crop yields, particularly in years of adverse weather. The soil is weakly structured and susceptible to surface capping by heavy rain or mechanical damage if worked when wet. Samples were collected from the top 10 cms of soil from two adjacent fields with contrasting cultivation histories. Characteristics of the soils are shown in Table 1, which indicates the lower organic content associated with cultivation.

Air dried samples of Bromyard 1 soil were gently sieved to obtain 2–4 mm aggregates in the natural state, fully aged and therefore physically stable. These were moistened to the equivalent plastic limit water content with either:-

(a)   deionised water
(b)   a solution of vancomycin (100 mg/g soil) (Sigma Chemical Co., Ltd.) as bactericide.
(c)   a solution of cycloheximide (100 mg/g soil) (Aldrich Chemical Co., Inc.) as fungicide.
(d)   a solution containing sodium azide (0.5 mg/g soil) and mercuric chloride (0.5 mg/g soil) as a general sterilant. (Tisdall et al., 1978).

The samples were incubated at 20°C and growth of fungi and bacteria, and corresponding aggregate stabilities were determined at 3 or 5 day intervals for 30 days. The appearance of samples during this time was examined using an ISI 60A scanning electron microscope. Microbial populations were determined by standard procedures, plating on Oxoid nutrient agar. Fungal growth was measured by HPLC determination of ergosterol (Seitz et al., 1977). Aggregate stability was determined by a turbidimetric method (Williams et al., 1966).

Organic matter, in particular the polysaccharide fraction of the soil, was degraded by treating soil samples at 20°C with 0.05 M sodium chloride for 48 hours, followed by 0.05 M sodium periodate for 48 hours, and finally by 0.1 M sodium borate for 48 hours (Clapp & Emerson, 1965a and b; Cheshire *et al.*, 1983).

Changes in aggregate stability after simulated cultivations were studied using air-dried soil in which natural aggregates were destroyed by passing through a 0.25 mm sieve. The pulverised soil was re-wetted with either deoinised water or with a sterilising solution. Soil was remoulded at plastic limit water content, allowed to equilibrate overnight, remoulded again, and then rolled into balls of 10–20 mm diameter. Stability of subsamples cut from these (8 replicates) was determined turbidimetrically as before.

The change in matric water potential which accompanies the reorientation of clay particles in soil during the process of "age-hardening" (Utomo & Dexter, 1981), was measured by means of the tensiometer shown in Figure 1. (For details, see Molope *et al.*, 1985).

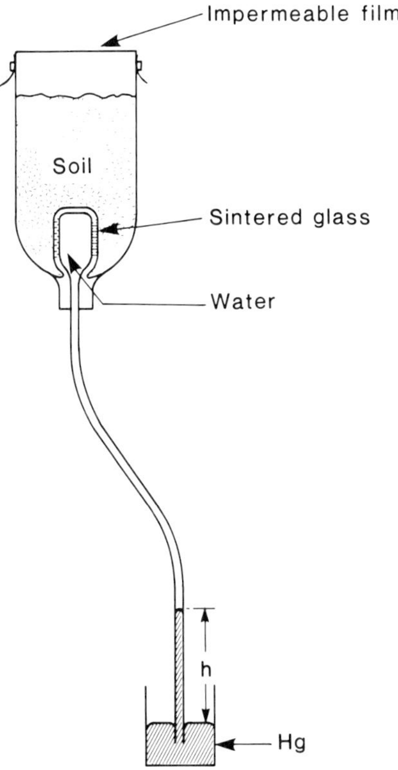

FIGURE 1   Measurement of matric water potential.

RESULTS AND DISCUSSION

A. Results with natural physically stable aggregates

Figure 2 shows the increase in aggregate stability over the first 6 days after moistening with deoinised water. This initial increase in stability occurred in parallel with visible fungal growth. Thereafter stability declined, accompanied by visible deterioration of the fungal mycelia. Stability was greatly decreased by treatment with periodate, demonstrating the prominent role of polysaccharides throughout the period of incubation but the rise in stability

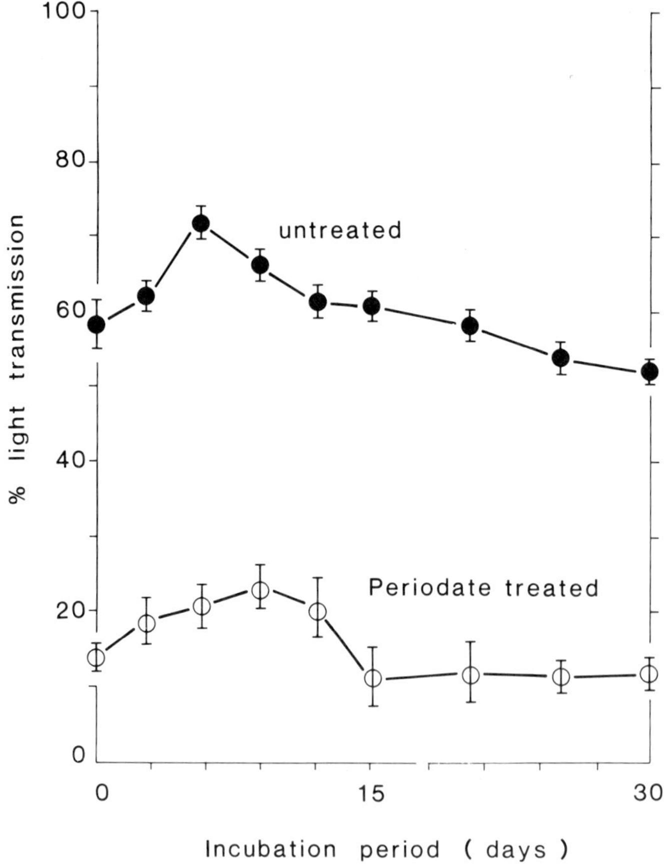

FIGURE 2   Change of water stability of natural aggregates of a physically stable unsterilised soil incubated at 20°C after moistening to plastic limit, and relative stability of similar samples after periodate treatment (see text).

Plate 1:
a   Sterilised physically stable soil. X800.
b   Initial growth of fungal hyphae (after 3 days) on soil moistened with deionised water. Incubated at 20°C. X400.
c   Development of fungal growth after 6 days. X600.
d   Matured hyphae, sporangia and spores after 9 days growth. X300.

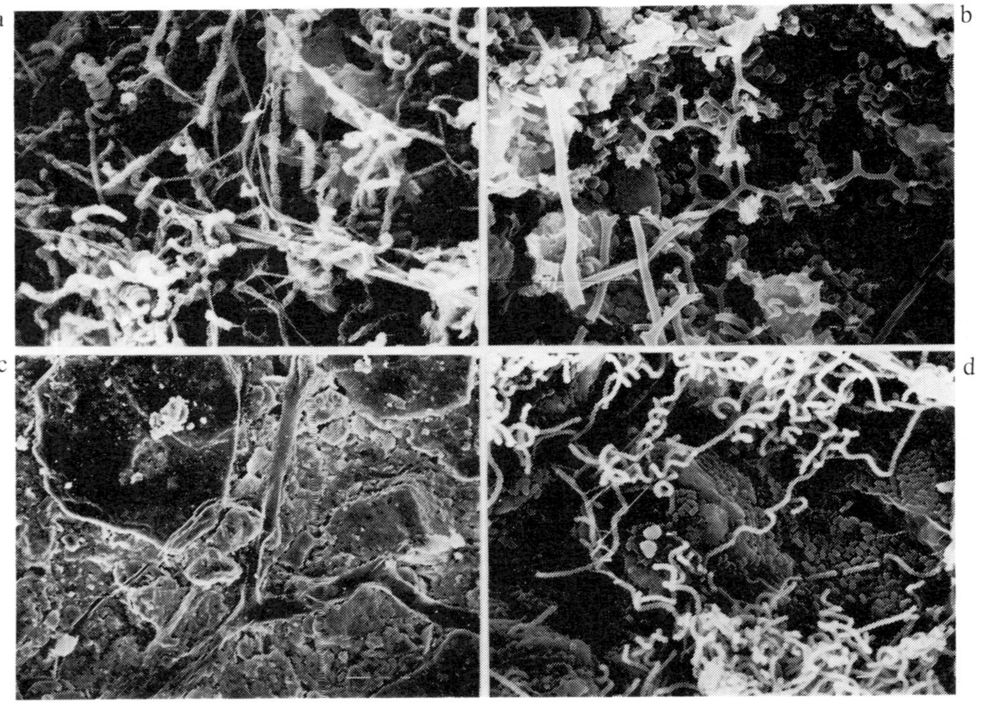

Plate 2:
a   Fungi, bacteria and actinomycetes after 15 days incubation. X1900.
b   Damage to hyphae by unknown grazing organism (25 days incubation). X1400.
c   Ruptured and empty fungal hyphae (30 days incubation). X340.
d   Disappearance of fungal and actinomycetal hyphae after 30 days incubation, and appearance of spores and conidia. X1400.

after periodate treatment over the first 12 days shows the independent aggregating function of fungal mycelia and their resistance to periodate degradation.

Some of the biological developments were revealed by the SEM pictures reproduced in Plate 1. Plate 1a, a picture of sterilised control soil after 3 days incubation at 20°C, shows the soil mineral particles and their orientation. After 3 days incubation at the plastic limit moisture content unsterilised soil exhibited some fungal growth (Plate 1b). After 6 days (Plate 1c) there was prolific growth of fungal hyphae, forming a reticulum of threads conferring maximum stability on the soil aggregates (cf. Fig. 2). In Plate 1d (9 days incubation) various sporangia and spores can be seen. These might germinate

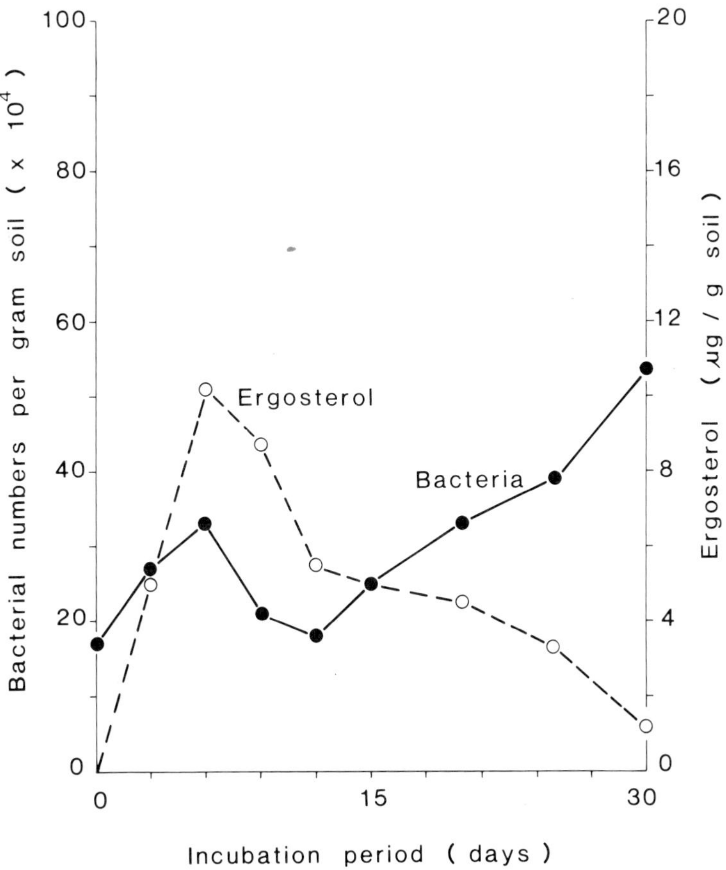

FIGURE 3   Bacterial numbers and a measure of fungal cell wall material during incubation at 20°C of untreated Bromyard soil.

or remain dormant depending on the availability of readily usable substrate. Some hyphae have already begun to deteriorate, either by the action of extracellular enzymes liberated by other organisms or by autolysis (Lloyd & Lockwood, 1966).

Plate 2a shows the proliferation of fungi, bacteria and actinomycetes after 15 days. Actinomycetes are able to attack fungal hyphae by means of extracellular chitinases. Damage to hyphae by some form of grazing organism at 25 days as shown in Plate 2b, while disrupted hyphae with damaged cell walls are shown in Plate 2c (30 days). Both in Plate 2c and 2d a noticeable reduction in visible organisms has occurred. In Plate 2d (30 days) both fungal and actinomycetal hyphae have disappeared. Spores and actinomycete

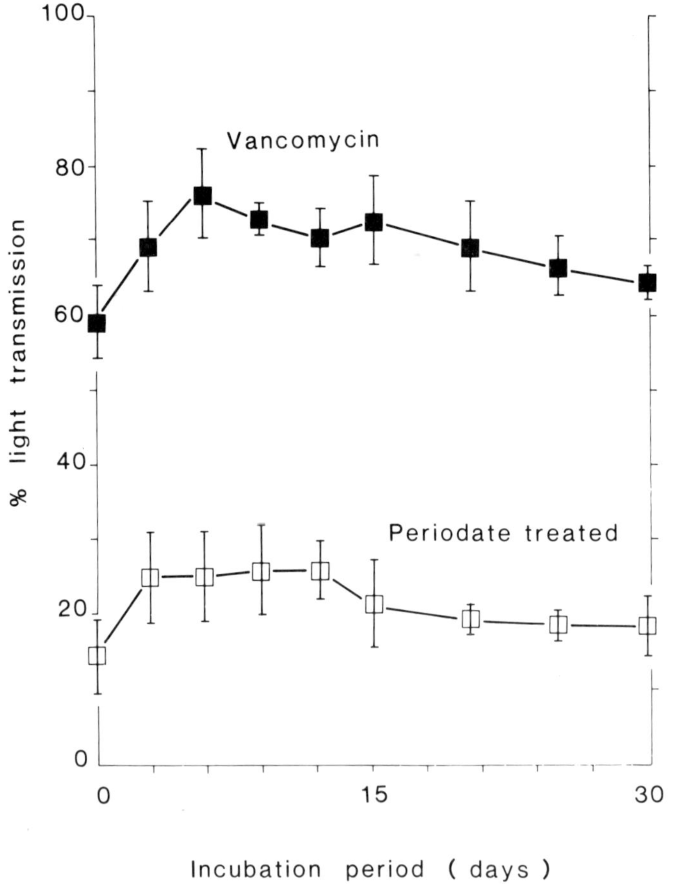

FIGURE 4   Aggregate stability of soil incubated with vancomycin to inhibit bacterial growth, and similar soil treated with periodate after incubation with vancomycin.

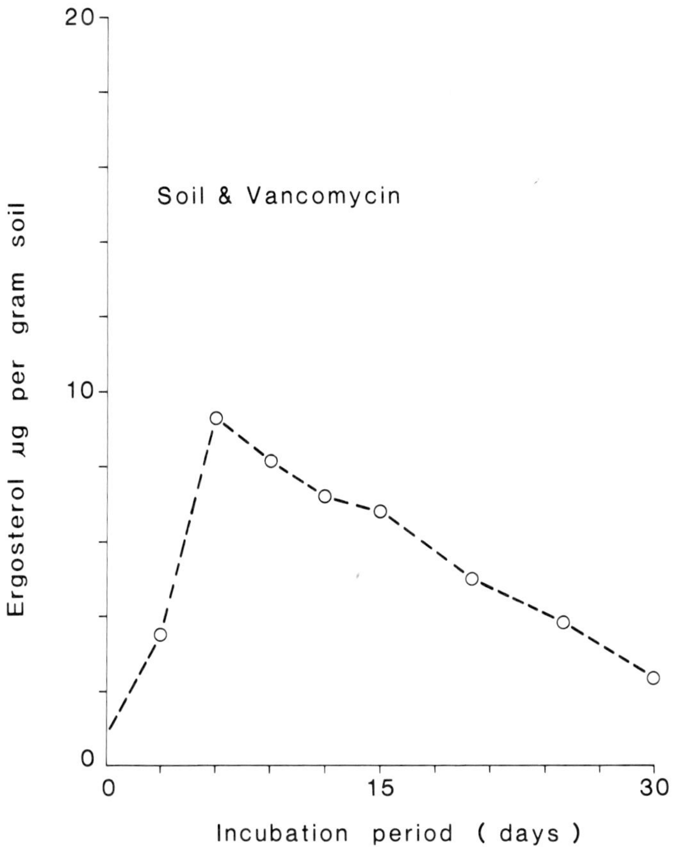

FIGURE 5   Ergosterol measurement of fungal cell wall material in soil treated with vancomycin.

conidia are visible, which will remain dormant until suitable conditions, including a renewed supply of substrate, occur.

The impressions gained from the SEM pictures were refined by the measurements made of bacterial numbers and fungal growth. Figure 3 gives the results for untreated Bromyard soil at a moisture content equal to the plastic limit. Fungal growth (measured as ergosterol) reached a maximum after the first 6 days and declined thereafter. (Ergostorol is the major sterol component of fungal cell walls, and is more reliable as a measure of fungal growth than chitin, which is present in soil in varying amounts in the form of insect and arthropod exoskeletons).

The decline in fungal cell wall material may be attributed to lysis by bacteria and actinomycetes, which proliferate at the expense of fungi, or to fungal

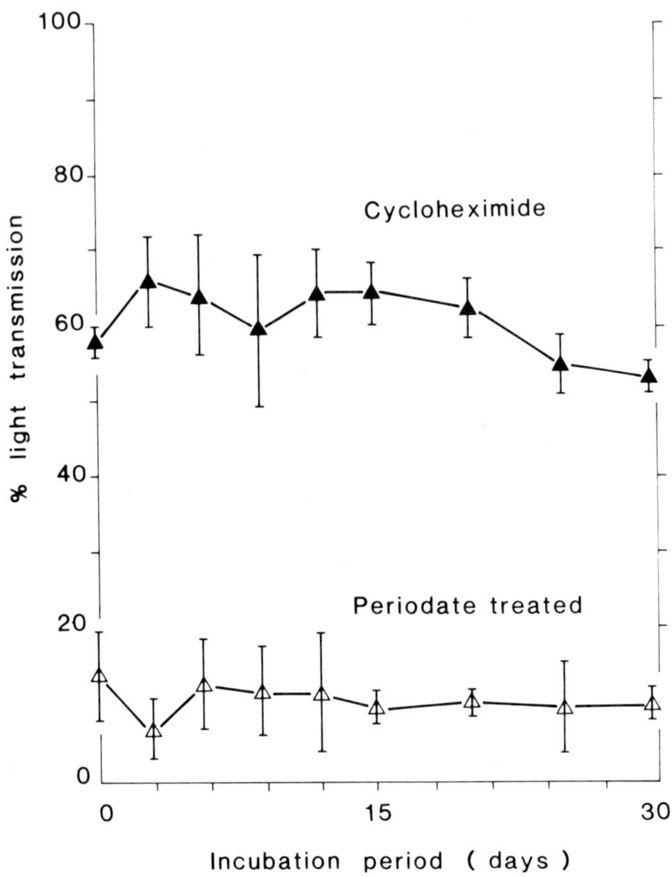

FIGURE 6   Aggregate stability of soil incubated with cycloheximide to inhibit fungal growth.

autolysis. Bacterial numbers continued to increase while fungal decline was in progress.

Comparison of Figures 2 and 3 shows that maximum aggregate stability coincided with maximal fungal growth, but that decline in aggregate stability did not keep pace with fungal decline. The greater part of the constant component of aggregate stability was destroyed by periodate treatment, indicating that it was probably attributable to polysaccharide materials.

When the soil was treated with vancomycin to inhibit bacterial growth, aggregate stability increased as before over the first 6 days of incubation (Fig. 4). The periodate resistant component increased in size compared with Figure 2, when bacteria were present, and did not decline to the same extent after 9 days. The fungal mass, as measured by ergosterol (Fig. 5) did not decline as quickly from the 6 day maximum as it had in the presence of bacteria, but there was a decline (the absence of bacteria was checked by plating, but is not shown in Figure 5) which would be attributable either to autolysis or to attack by actinomyetes.

When fungal growth was inhibited by cycloheximide, aggregate stability (Fig. 6) did not increase during the first 6 days to the same extent as in untreated or vancomycin treated soil. Aggregate stability changed little during the 30 days of incubation, and the final stability was much the same as in untreated soil after fungal hyphae had disappeared. After periodate treatment the aggregate stability of cycloheximide treated soil remained uniformly low, showing that in the absence of binding hyphae the main aggregating substances were polysaccharides.

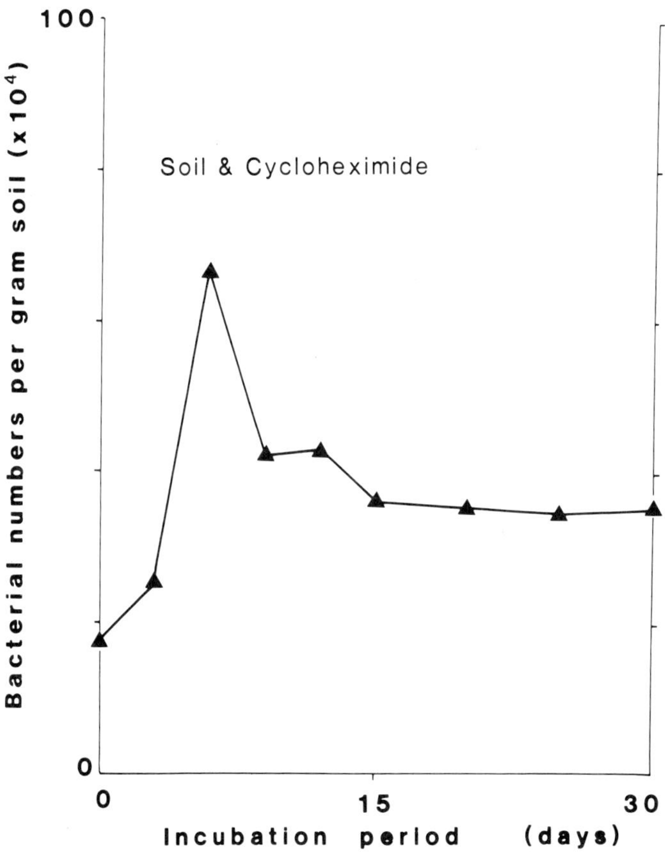

FIGURE 7 Bacterial numbers during incubation of soil treated with cycloheximide.

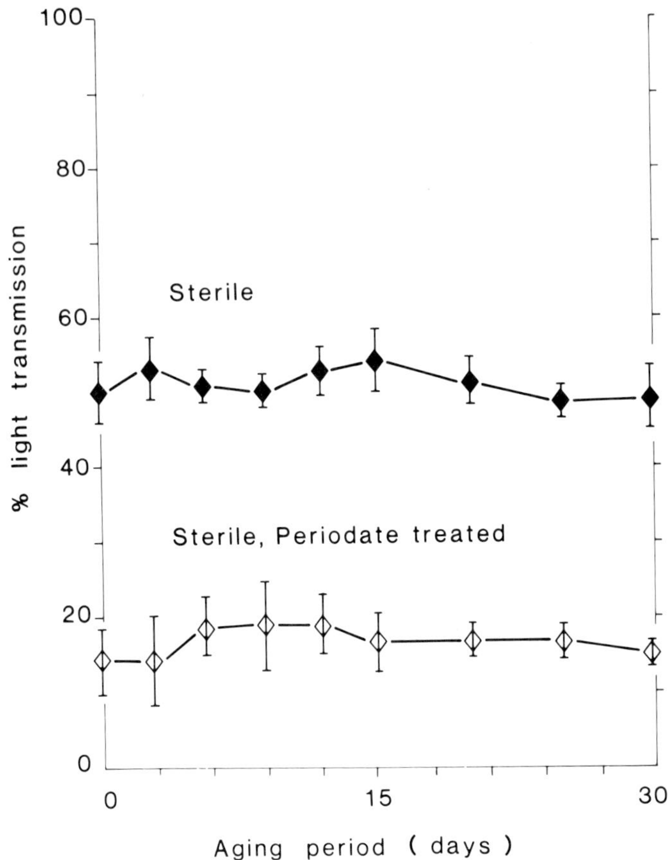

FIGURE 8   Aggregate stability of soil sterilised with sodium azide and mercuric chloride.

Figure 7 shows bacterial populations during incubation in the presence of cycloheximide (ergosterol estimations checked the absence of fungi). A comparison with Figure 3 shows the much more rapid increase of bacterial numbers in the absence of fungal competition. After reaching a peak at 6 days, numbers declined to about half the maximum and remained fairly steady after about 15 days.

When soil was sterilised with sodium azide and mercuric chloride neither fungal nor bacterial growth took place. Aggregate stability measured over an aging period equal in length to the previous periods of incubation revealed no significant changes (Fig. 8). Treatment with periodate removed most of the stabilising substances, presumed to be polysaccharides, leaving a residual aggregate stability which similarly remained constant over the aging period.

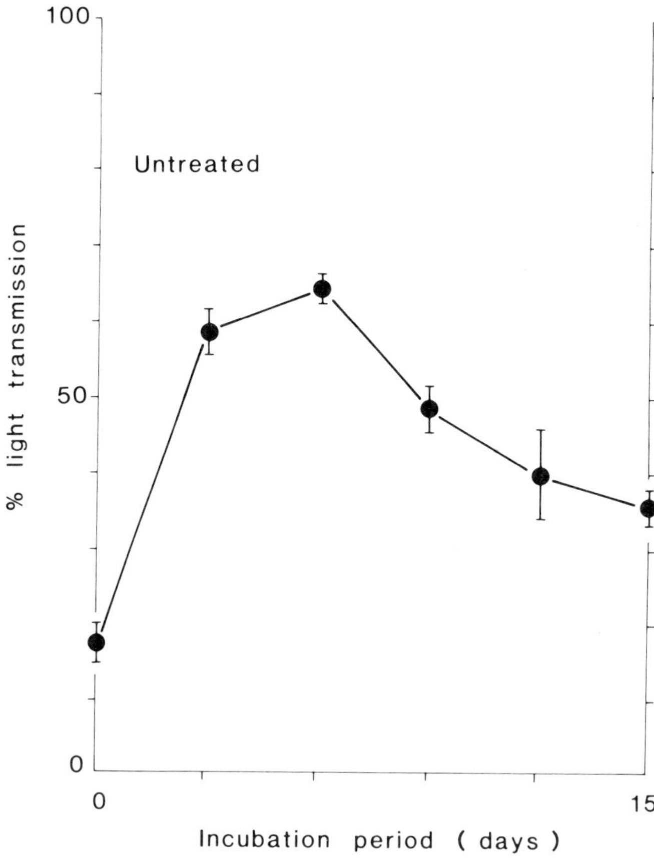

FIGURE 9   Changes in aggregate stability of pulverised soil moistened and remoulded to simulate cultivations.

Soil which was originally in a stable physical state thus kept its initial aggregate stability unchanged when biological activity was prevented during aging.

B. Results with freshly moulded aggregates

Newly moulded aggregates, formed by processes simulating vigorous cultivations, had very low aggregate stability immediately after pulverised soil had been re-moulded. Figure 9 shows the rapid increase in aggregate stability during incubation of moistened soil which had no added inhibitors of biological activity, and shows that the initial level of stability was much lower

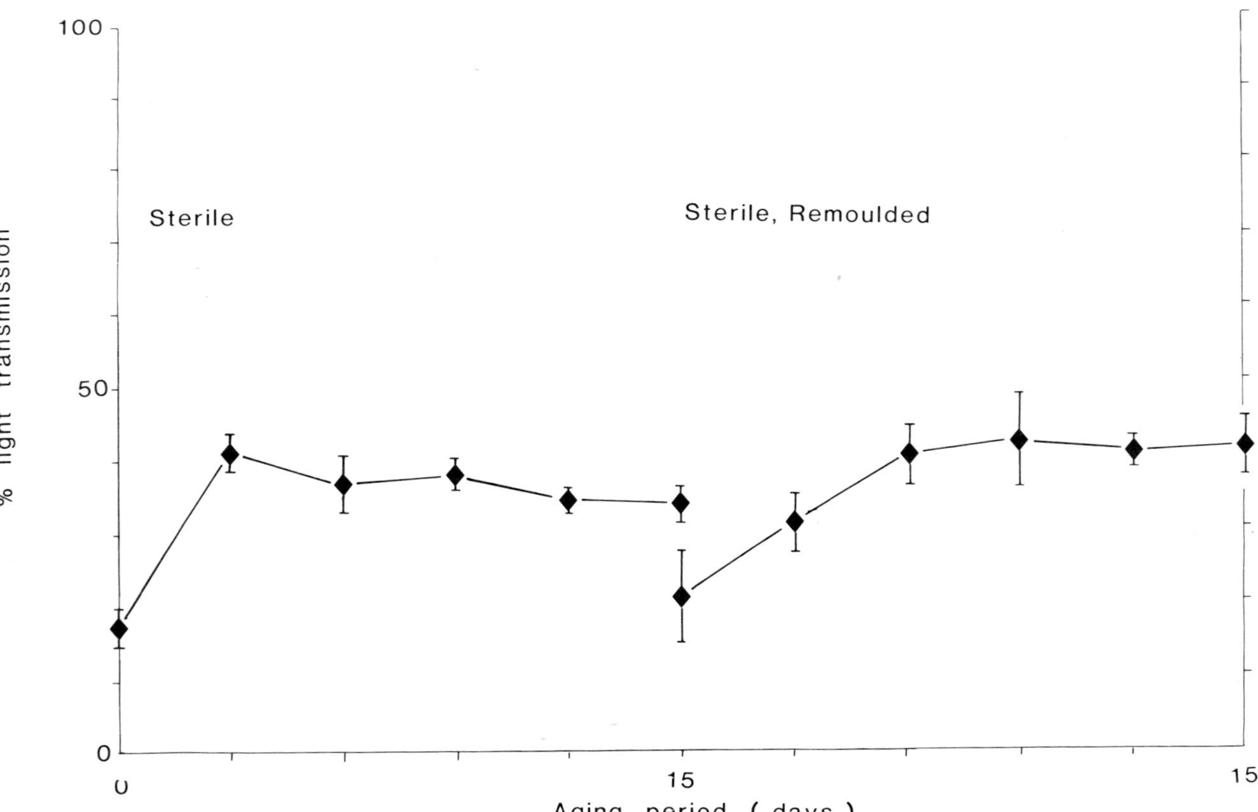

FIGURE 10   Changes in aggregate stability during aging of sterile artificial aggregates of remoulded soil.

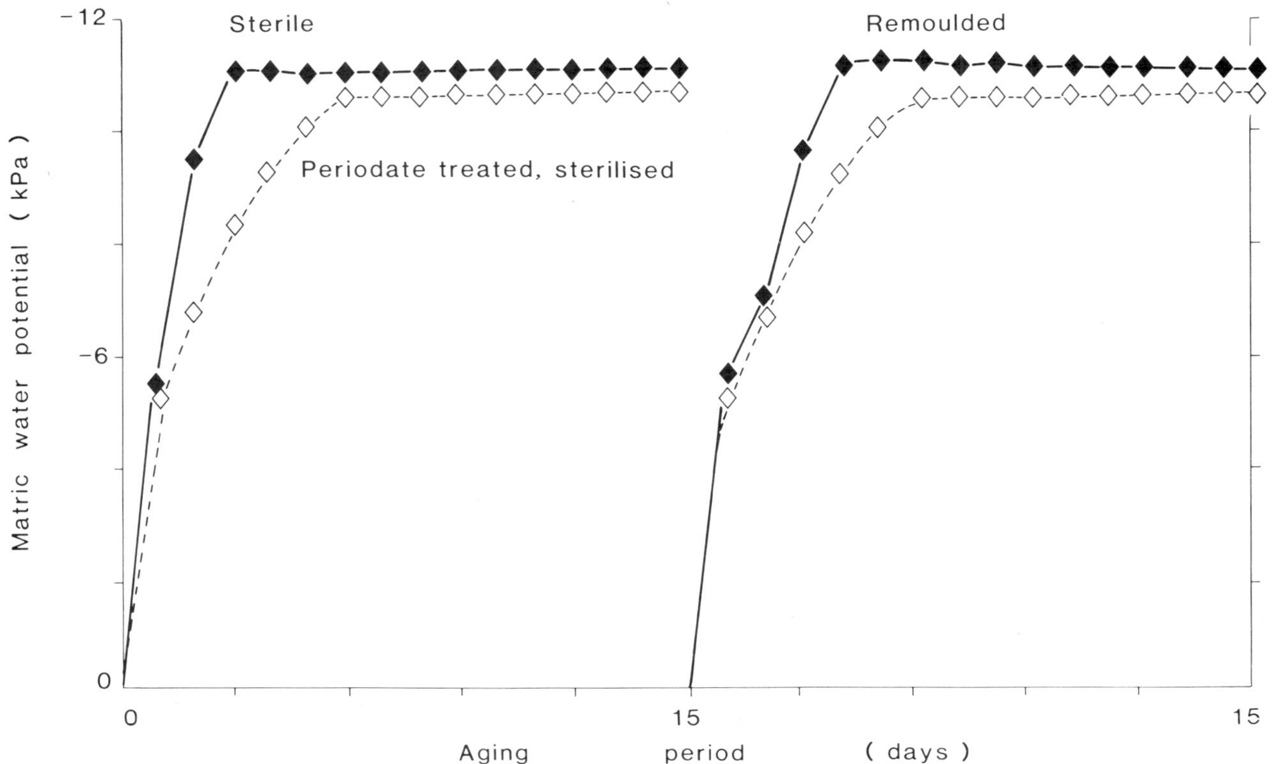

FIGURE 11 Changes in soil matric potential during aging of sterile remoulded soil.

than that of natural aggregates of the physically stable soil (Cf. Fig. 2). Biological changes would account for some of the increase in aggregate stability exhibited in Figure 9, but a large component of the change was purely physical, as is shown in Figure 10, which portrays the increase in aggregate stability in sterilised soil. Stability increases during the first 3 days, then remains constant. At 15 days the soil was remoulded. There was immediately a fall in stability, then recovery over 6 days to the previous equilibrium stability, which thereafter remained constant. The process took place at constant temperature, was time dependent, and reversible. These were the criteria for thixotropic change proposed by Mitchell (1960). He explained thixotropic behaviour in clays by hypothesising that plate-like clay particles become orientated by external shearing forces into more uniformly parallel arrangements. This would occur during cultivation, and aggregate stability would be minimal, because the clay platelets would be able to slip past each other. During the aging process, thermal oscillations cause the platelets to become randomly orientated, and the rigidity imparted to the aggregates would result in increased stability.

The increase in randomisation of the clay particles during thixotropic change is accompanied by increasing soil moisture tension (i.e. metric water potential becomes more negative). The changes in matric potential during aging of sterilised Bromyard soil are shown in Figure 11. It is seen that these changes reflect the changes in aggregate stability depicted in Figure 10, and are reversible in the same way. Changes in soil water matric potential are very little affected by periodate treatment, in contrast to the large effect on aggregate stability.

There are some implications for soil management practice in these findings that both physical and biological processes are involved in the development of soil aggregate stability. The equilibrium stability of stable soil aggregates results in part from the achievement of a thixotropic gel state and in part from a dynamic equilibrium of biological changes involving both structures and metabolic products. Cultivation disturbs the stability of aggregates. The improvement of stability after cultivation can be a deception, in that what is observed is either a return to an equilibrium physical stability, or is caused by a flush of fungal and bacterial growth which must deplete the stored organic matter capital of the soil.

An understanding of the mechanisms involved in the structural deterioration imposed by cultivations and the restoration of structure under pasture or other farming systems must form the basis for sound and scientific management of soil in sustainable agriculture.

SUMMARY

Results from studies of a Bromyard soil, typical of those obtained from several

English and Scottish soils, are reported, which examine biological and physical processes involved in the development of soil aggregate stability.

The initial increase in aggregate stability of incubated soil was accompanied by prolific fungal growth. This was clearly seen by electron scanning microscope examination, and was estimated by means of ergosterol measurements. The increase in aggregate stability may be partly explained by retention of elementary soil particles within a reticulum of fungal hyphae. At the same time an increase of bacterial populations was demonstrated by means of plate counts. Bacterial effects on aggregate stability appear to be associated with polysaccharide products of bacterial metabolism, and were investigated by means of periodate oxidation.

Growth of fungi and bacteria were inhibited respectively by cycloheximide or by vancomycin to demonstrate the separate effects. When all biological activity was prevented by using sodium azide and mercuric chloride a residual process remained which still caused some increase of stability in aggregates formed by simulated cultivation of pulverised soil. This thixotropic change was shown to be reversible and to be associated with an increase in soil moisture tension; it may therefore be explained by orientation and later disorientation of clay particles in the soil aggregates.

References

Cheshire, M.V., Sparling, G.P. & Mundie, C.M. (1983). Effect of periodate treatment of soil on carbohydrate constituents and soil aggregation. *Journal of Soil Science,* **34**, 105–112.

Clapp, C.E. & Emerson, W.W. (1965a). The effect of periodate oxidation on the strength of soil crumbs. I. Quantitative studies. *Proceedings Soil Science Society of America,* **29**, 127–130.

Clapp, C.E. & Emerson, W.W. (1965b). The effect of periodate oxidation on the strength of soil crumbs. II. Qualitative studies. *Proceedings Soil Science Society of America,* **29**, 130–134.

Lloyd, A.B. & Lockwood, J.L. (1966). Lysis of fungal hyphae in soil and its possible relation to autolysis. *Phytopathology,* **56**, 595–602.

Mitchell, J.K. (1960). Fundamental aspects of thixotropy in soils. *Journal of the Soil Mechanics and Foundations Division, American Society of Civil Engineers,* **86**, SM3, 19–52.

Molope, M.B., Grieve, I.C. & Page, E.R. (1985). Thixotropic changes in the stability of molded soil aggregates. *Soil Science Society of America Journal,* **49**, 979–983.

Seitz, L.M., Mohr, H.E., Burroughs, R. & Sauer, D.B. (1977). Ergosterol as an indicator of fungal invasion of grains. *Cereal Chemistry,* **54**, 1207–1217.

Tisdall, J.M., Cockcroft, B. & Uren, N.C. (1978). The stability of soil aggregates as affected by organic materials, microbial activity and physical disruption. *Australian Journal of Soil Research,* **16**, 9–17.

Utomo, W.H. & Dexter, A.R. (1981). Age hardening of agricultural top soils. *Journal of Soil Science,* **32**, 335–350.

Williams, B.G., Greenland, D.J., Lindstrom, G.R. & Quirk, J.P. (1966). Techniques for the determination of the stability of soil aggregates. *Soil Science,* **101**, 157–163.

Alley Cropping: Trees as Sources of Green-Manure and Mulch in the Tropics

G.F. Wilson, B.T. Kang and K. Mulongoy[1]

International Institute of Tropical Agriculture, PMB 5320, Ibadan, Nigeria

INTRODUCTION

In the development of technologies for improving the efficiency of agriculture, it must be remembered that agricultural systems are ecosystems in which man plays a dominant role in directing the energy flow into food and other material necessary for his survival. Therefore, the efficiency of an agricultural system is determined, to a certain extent, by the level of man's interference or management, and his success in diverting energy in the desired direction. Many tropical systems are inefficient not because of natural resource limitation but because of limited management. For example, a bush fallow system in which the rotation consists of two years cropping followed by eight years naturally regenerated fallow, has management limited to the cropping period or only 20% of the cycle. To correct this discrepancy some form of fallow management must be developed to ensure that more of the energy generated during that period is directed to meeting man's needs.

Farmers in the tropics have long realized the efficiency of some trees in restoring soil fertility, and have endeavoured to give these plants selective advantages. In Africa, species such as *Acioa bateril*, *Macrothylum* sp., *Anthonatha macrophylla* and *Gliricidia sepium* are encouraged and protected during cropping and given a better chance of dominating the fallow (Benneh, 1971; Wilson & Kang, 1981). In Asia, farmers use *Leucaena leucocephala* and *Sesbania grandiflora* in the fallow (NAS, 1979).

It would be unfair to say that researchers have not tried to find suitable fallow species for the humid tropics, but the results have not contributed significantly to food production in the region. The major short-coming has been the emphasis on herbacious species in environments dominated by trees and shrubs. Milsum & Bunting (1928), in suggesting that shrub legumes may

---

[1]Agronomist, Soil Fertility Scientist and Soil Microbiologist respectively.

be more suitable than herbacious legumes for soil restoration in the humid tropics, pointed out that nitrogen released from decomposing plant tissue could be more important than that exuded from roots in these environments. By taking the above suggestion even further we are now in an era where trees as well as woody shrubs are being hailed as the saviours of the tropics, and are receiving much attention in the drive to improve or develop more efficient alternatives to the bush fallow.

Bush Fallow

Ruthenberg (1971) described bush fallow as an agricultural system in which a few years of cultivation alternate with many years of resting or fallow during which soil fertility is restored. During the fallow period, the land is colonized, in succession, by various plant species. These plants are the active biological entitities that rejuvenate the upper soil that was depleted during cropping. Nutrients absorbed by them, especially deep rooted woody perennials, from the lower soil layers and the atmosphere through fixation are deposited on the soil surface through leaf fall and at bush clearing. The duration of the fallow is critical to the productivity of the system, and depends on the fertility of the soil parent material, the level of top-soil depletion at the end of cultivation, the plant species in the fallow and the soil organisms, including those interacting with plants in the fallow and free living ones that influence litter decomposition and humus formation.

Of the above factors, the plant species comprising the fallow appear to be the most important. Some species, especially legumes, are capable of hastening soil enrichment by fixing atmospheric nitrogen through symbiotic interaction with *Rhizobium* bacteria. In most bush fallow systems, plant succession following cultivation follows a random pattern in which there is no guarantee that the more efficient soil restoring species will be the most abundant or dominant ones. Competition from the less efficient ones may also reduce the effectiveness of the efficient ones. It is believed that the soil fertility restoring rate of bush could be improved if it was organized to allow selected effective soil restoring species to dominate (Wilson & Kang, 1981).

Agroforestry

During the 1970's, there has been increased awareness of the role of trees in tropical ecologies (Bene *et al.,* 1977; NAS, 1977, 1979; Sanger, 1977; De las Salas, 1979; Mongi & Huxley, 1979; Combe *et al.*, 1981; Getahun *et al.*, 1982; MacDonald, 1982; Huxley, 1983). The fact was emphasized by the droughts that swept the Sahelian region of Africa, where humans and cattle perished or fled from their famine ravaged homeland. Here man in his effort to provide more land for growing crops and rearing animals, destroyed the trees that were important in recycling water and other essential things. Paradoxically, in the effort to survive, those people unwitting destroyed that which was essential for their survival.

How then can the conflict between man and trees be resolved so that both can survive and flourish together? There are no ready answers, but out of that tragedy grew the concept of 'Agroforestry', a branch of science dedicated to the integration of crop and livestock production with forestry activities to improve or prevent further degradation of fragile ecosystems, especially those of the tropics (Wassink, 1977; Poulsen, 1978; King, 1979).

Numerous agroforestry concepts aimed at increasing food production while retaining trees in a stable ecosystem have been developed (Mongi & Huxley, 1979; Huxley, 1983). In nearly all these systems nutrients released through leaf decomposition are regarded as the most important factors in soil fertility restoration. These leaves can be incorporated into the soil as green manure or retained on the surface as mulch. The two methods of leaf utilization are important considerations for effective implementation of alley cropping (Wilson & Kang, 1981; Kang *et al.*, 1981a, 1981b; Sumberg, 1984) one of the most promising agroforestry concepts for the tropics.

ALLEY CROPPING

Alley cropping (Figs. 1 & 2) is the growing of crops, usually food crops, in alleys formed by trees or woody shrubs that are cut back at crop planting and maintained as hedge-rows by frequent trimming during cropping (Wilson & Kang, 1981). The leaves and twigs from the cut trees are added to the soil as green manure (Fig. 3) or mulch (Fig. 4). The system was developed as an alternative to bush fallow and is regarded by some as an improved bush fallow. Improving the bush fallow is a new concept that has strong implications, as it aims at increasing the efficiency of a system that, through independent evolution in various parts of the tropics, shows good compatibility with that environment. As in the bush fallow, alley cropping depends on nutrient recycling through decomposition of leaves from deep rooted and nitrogen fixing trees. However, random succession and plant arrangement is removed and one or more selected species arranged in an ordered pattern (hedge-rows and alleys) are allowed to dominate the fallow.

Because the trees are kept alive during cropping the land is always under the influence of the selected fallow species which, by dominating even short fallows, restores soil fertility rapidly.

A major feature of alley cropping is the ability of the deep-rooted species to absorb from the lower soil strata soil moisture not available to annual crops.

FIGURE 1   Maize in a 4 metre wide alley formed by *Leucena leucocephala*.

FIGURE 2   Maize in a 2 metre wide alley formed by *L. leucocephala*. Note the mulch formed by twigs.

FIGURE 3   Incorporating green manure from leaves of *Gliricidia sepium*.

FIGURE 4   Cassava mulched with leaves and twigs from *G. sepium*.

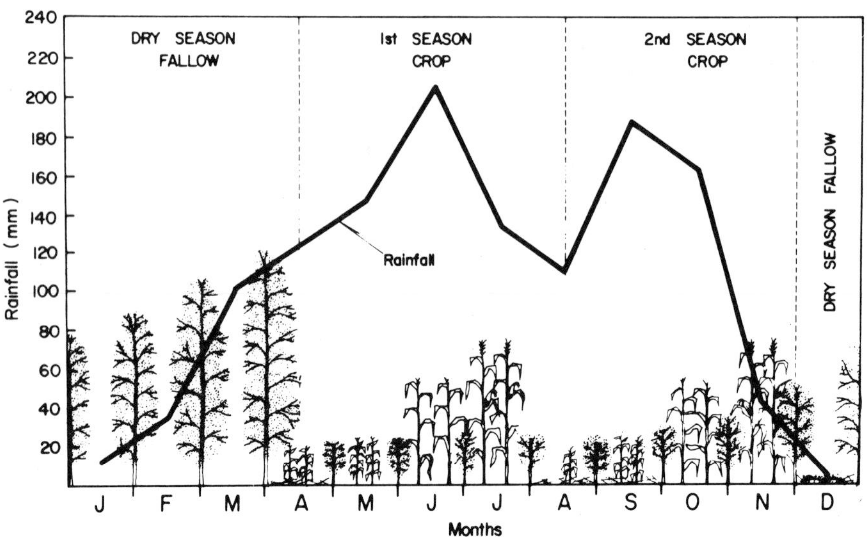

FIGURE 5   Alley cropping management with two cropping seasons and dry season fallow.

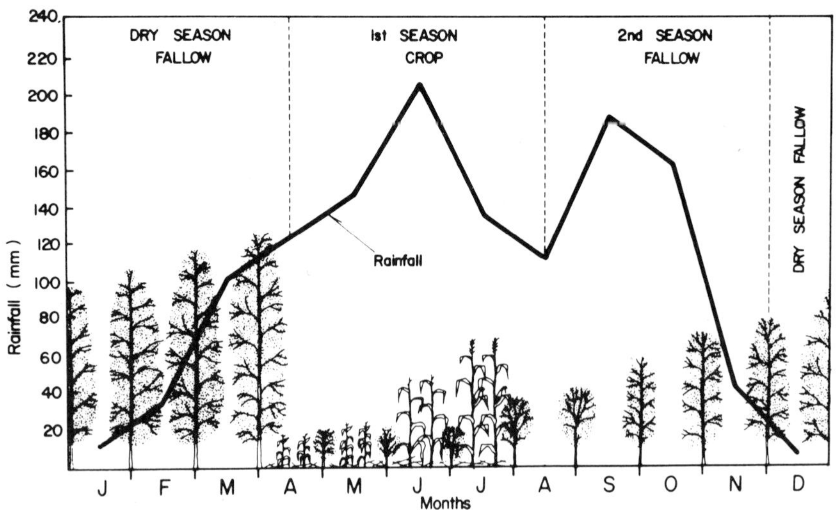

FIGURE 6   Alley cropping management with first season cropping and dry season fallow.

This moisture allows them to remain functional in the dry season, using the abundant radiant energy of this season to run a biological factory that recycles nutrients, fixes nitrogen (when legumes are used), synthesizes organic matter, pumps water into the atmosphere, and shades the soil providing a more favourable environment for the activities of beneficial soil organisms. Kowal & Kassam (1978) estimate that incoming radiation during the dry season in the Guinea savanna of West Africa ranges between 49,000 and 142,000 cal.cm^{-2}. Normally all this energy goes to waste, but with alley cropping much would be captured by the trees in the fallow. Two alley

TABLE 1

Nutrient composition of prunings of four tree and shrub species grown on Egbeda sandy loam (Oxic Paleustalf) (Koudoro, 1982)

Species	N	P	K	Ca	Mg
			(%)		
Gliricidia sepium	4.21	0.29	3.43	1.40	0.40
Leucaena leucocephala	4.33	0.28	2.50	1.49	0.36
Alchornea cordifolia	3.29	0.23	1.74	0.46	0.20
Acioa barterri	2.57	0.16	1.78	0.90	0.27

TABLE 2

Effect of applications of leucaena prunings and inorganic nitrogen on grain yield of maize grown on Apomu loamy sand (Psammentic Ustorthent) (Kang *et al.*, 1981a)

Leucaena rate (tons/ha)	N rate (kg N/ha)	Leucaena prunings	
		Incorporated	Mulch
		(grain yield/kg/ha)	
0	0	1283	1740
	50	2093	2218
	100	3315	3138
5	0	2313	2013
	50	3035	2300
	100	3453	3028
10	0	3213	1855
	50	2578	2338
	100	3068	3023
Mean		2705	2406

LSD .05 — Between means of methods of leucaena placement, 688.
Between treatments in different leucaena placement methods, 1146.

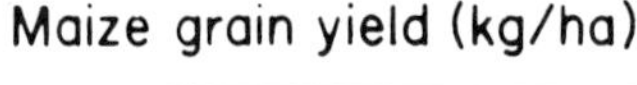

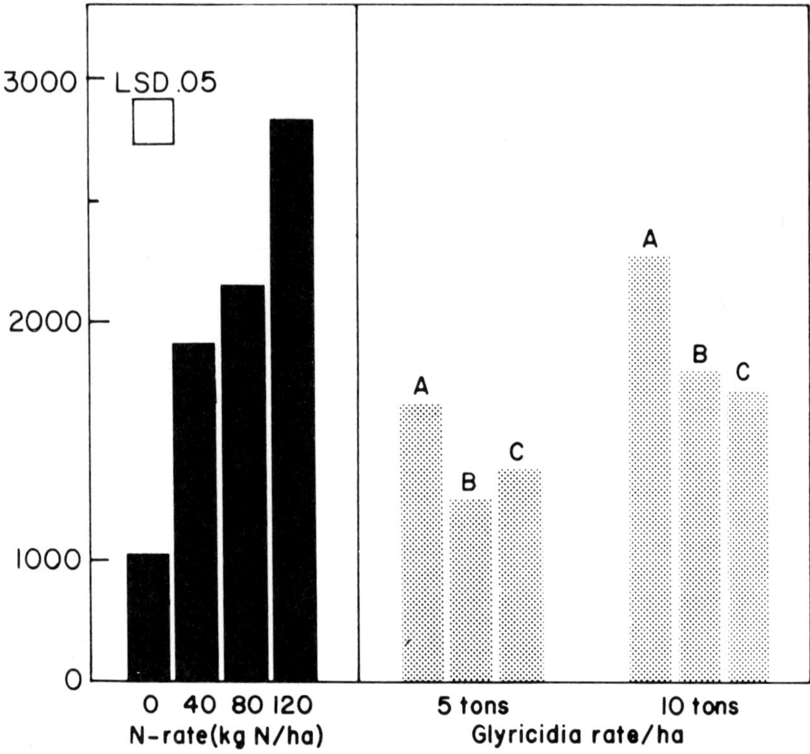

FIGURE 7   Effect of the application of nitrogen and *Gliricidia* tops on grain yield of maize grown on Apomu soil (Psammentic Usthorthent) at Ibadan. (*Gliricidia* top placement; A = band incorporated; B = band and applied at surface; C = surface broadcast).

cropping managements in which dry season fallow is used for soil restoration are shown in Figures 5 and 6. In the second illustration (Fig. 6) the short second rainy season is also used to extend the restoration period.

Leucaena (*Leucaena leucocephala*) and gliricidia (*Gliricidia sepium*), trees well known for nitrogen fixation and soil fertility restoration (Guevarra *et al.*, 1978; NAS 1979), were the main species used in the pioneering work on alley cropping. In Table 1 the nutrient contents of these legumes are compared with *Alchornea cordifolia* and *Acio barterii*, two non legume species common in bush fallow in the humid lowland tropics of Africa which are also being tested in alley cropping (Koudoro, 1982).

Maize grown in alley cropping with both leucaena (Table 2) and gliricidia (Fig. 7) responded significantly to addition of the tree leaves to the soil, as compared to treatments where the leaves were removed. When leucaena alley

TABLE 3

Main season grain yield of maize variety TZPB alley cropped with *Leucaena leucocephala* on Apomu loamy sand (Psammentic Ustorthent) as affected by application of leucaena prunings and nitrogen (B.T. Kang, unpublished data)

N-Rate (kg N/ha)	Leucaena prunings	Year 1979	1980	1981*	1982	1983
		--------------------------- (tons/ha ---------------------------				
0	Removed	-	1.04	0.48	0.61	0.26
0	Retained	2.09	1.91	1.21	2.10	1.92
80	Retained	3.54	3.26	1.89	2.91	3.16
LSD .05		0.36	0.31	0.29	0.44	0.79

*Maize crop seriously affected by drought during early growth.

cropping was tested over a five year period on the low fertility Apomu loamy sand (Psammentic Ustorthent), maize yield declined in the control where leucaena leaves were not applied to the soil, but was maintained and was significantly higher than the control where the leaves were applied (Table 3).

Green Manure or Mulch

In the early days of agricultural research in the tropics much effort was devoted to herbaceous legume—green manure systems similar to those popular in temperate regions (Milsum & Bunting, 1928; Webster, 1938; Vine, 1953). Beneficial effects were shown but the method was not accepted widely. That farmers may be averse to green manure crops that occupy the land for the rainy season without giving direct return (Webster & Wilson, 1980) is an acceptable reason for their failure to accept green manuring in the tropics. However, the climatic constraints of most tropical areas may be more important factors. Except in the high rainfall forest zones, the main cropping seasons are proceeded by distinct, often severe dry periods that most herbacious cover crops do not survive. The dry residue formed when the cover crops die is often destroyed by fire. Should it escape the fire nothing but dry material with a high C:N ratio, which causes nitrogen tie-up, adversely affecting early crop growth and yield, is left. In fact, herbacious material suitable for green manure is rarely found in the derived savanna of humid, sub-humid and semi-arid tropics where cereals, the major beneficiaries of green manure, are grown. Even in Ibadan, with over 1,200mm rainfall, the severe dry season leaves *Mucuna utilis* as dry residue by the beginning of the major cropping season. Thus, an effective green manure as used by Vine (1953) would have used the major and most productive season for producing green manure for crops to be grown in the minor unpredictable and unreliable season. Such a system would be highly unacceptable to the subsistence farmer with strong aversion to risk.

Milsum & Bunting (1928) were among the first to point out that shrubs able to survive the dry season and retain a substantial amount of green leaves were the best potential sources of green manure for the tropics. They even suggested a cut and carry system in which the green manure from shrubs would be used in fields other than those in which it was produced. The results obtained with pigeon pea, a deep rooted shrub that is tolerant to drought, exemplifies the potential of this method (Webster & Wilson, 1980).

Trees and shrubs, unlike herbs, survive the dry season (Fig. 8) and usually have abundant leaves for green manuring at the start of the major rains. Even species such as *Gliricidia sepium*, that are deciduous if not pruned before the days shorten, present copious amounts of leaves by the start of the rains. Obviously leaf shedding is not in response to moisture stress, but to shortening day length which also triggers flowering.

Until the development of herbicides and consequently no-tillage, fallow residues were not usually considered a suitable mulch. The reason for this was that too much labour was required to remove the residue prior to tilling the land and then replacing the residue as mulch. However, once it was understood that tilling was not necessary, on some soils, for good yield if good

FIGURE 8   Fallowed land under *G. sepium* in the dry season. The trees continue growing and synthesizing organic matter that will be green manure or mulch for the next crop.

weed control could be maintained, then the cover crops that did not make good green manure were used as mulch (Wilson, 1978a, 1978b). As mulch, the effect was often opposite to that achieved with green manure on maize. Nitrogen deficiencies were noticed on maturing maize and yields were suppressed where nitrogen was not applied with an *in situ* mulch of *Mucuna utilis* (IITA, 1981; Akapa & Wilson, unpublished). However, with tomato, yields were not depressed by an *in situ* mulch of *Pueraria phaseoloides* nor were there significant responses to added nitrogen (Wilson, 1978a). Apparently, nitrogen release from the decomposing mulch was too slow for maize, which has a high early nitrogen demand, but not for tomato with a later demand peak.

The major advantages of trees are that they have deep roots which enable them to survive the dry season and to restore mineral nutrients to the surface soil; and by shading, which reduces soil temperature and evaporation, they produce a favourable environment for beneficial soil organisms, notably earthworms. Observation at IITA reveals that earthworm activities are higher under the shade of trees (alley cropping) than in soils that are not shaded. Consequently soils associated with alley cropping are usually of a good tilth suitable for no-till planting.

On loamy sand Apomu soil (Psammentic Ustorthent) gliricidia (Fig. 7) leaves were better applied as green manure than as mulch. Similar results were obtained with leucaena leaves. Read *et al.*, (1982), in pot trials with maize, observed that fresh leucaena leaves applied to pots to give 100 ppm N resulted

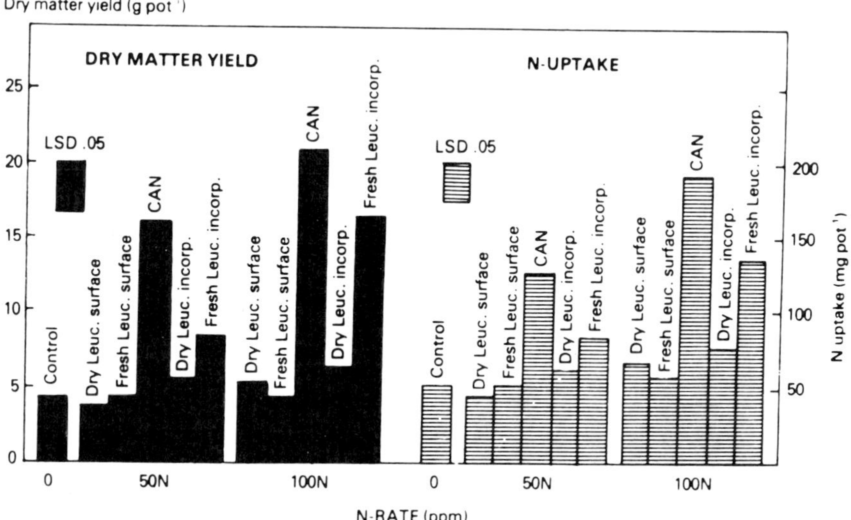

FIGURE 9   Effects of Fresh and dry leucaena as mulch and incorporated in soil. (CAN = Calcium ammonium nitrate)

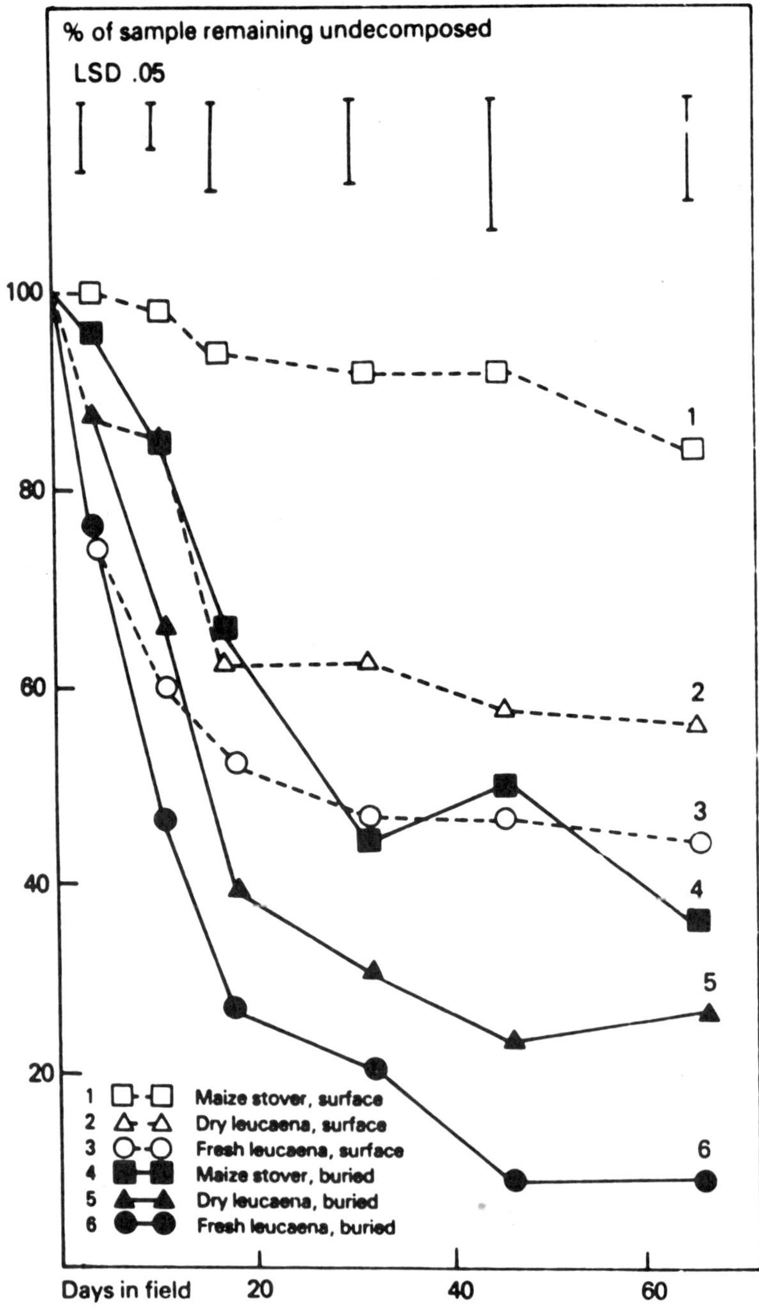

FIGURE 10   Decomposition of fresh and dry leucaena leaves as mulch and incorporated in soil.

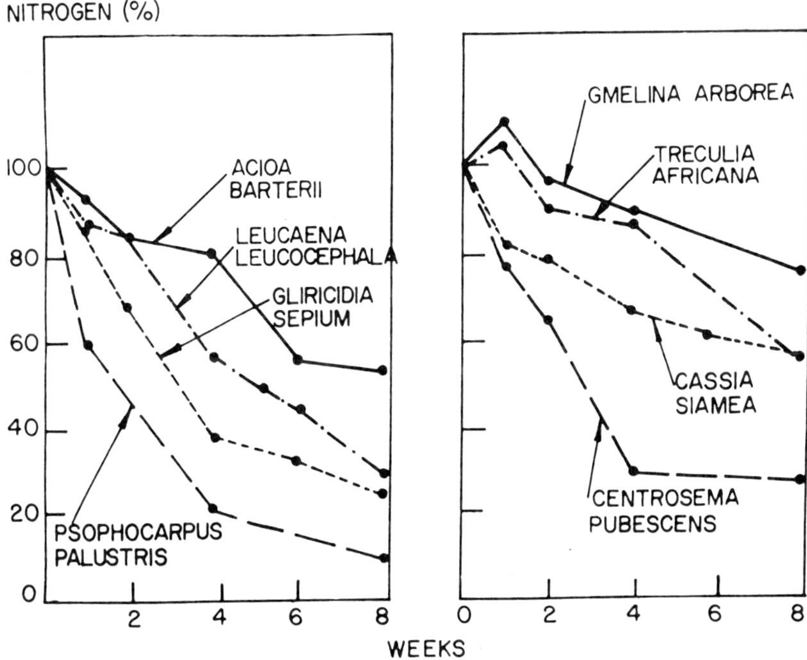

FIGURE 11 Nitrogen content (expressed as a percentage of the initial N content) of decomposing leaves of selected plants confined in nylon bags of 32 mesh size at Ibadan, Nigeria, 1982.

in significantly higher dry matter yield when they were incorporated as green manure than when retained as mulch (Fig. 9). When incorporated, fresh leucaena was better than dry leucaena at the high N rate of 100 ppm but not at 50 ppm. The overall trend suggested that incorporation was superior to surface application. The nitrogen uptake as influenced by the state of the leucaena leaves and the method of application are shown in Figure 9.

The superiority of incorporated leucaena was ascribed to more rapid decomposition and nutrient release (Fig. 10). In both fresh and dry states decomposition was faster when the leaves were buried. Rapid decomposition of buried leaves was due to the higher moisture content. Decomposition studies with leaves of various species in two environments showed that the decomposition rate was higher in the more humid Onne than at the dryer IITA environment (Table 4). Note that N loss was also more rapid under higher rainfall and was not proportional to the dry matter decomposition. Nitrogen loss was generally more rapid for the legumes than for the non legumes, *Acioa barterri* and *Gmelina Orborea* (Fig. 11). The exceptional legume was *Cassia siamea*.

TABLE 4

Field decomposition of leaves of selected plants at
IITA and Onne

Species	% dry matter after 10 weeks of decomposition		Percent N before decomposition	Number of weeks until 50% N loss	
	IITA	ONNE		IITA	ONNE
Acioa barteri	84	85	1.7	8.9	9.4
Centrosema pubescens	83	68	4.4	2.9	1.6
Cassia siamea	79	77	2.7	9.1	9.0
Gmelina arborea	78	70	2.2	<10	8.7
Tephrosia africana	7[b]	61	2.0	8.6	8.4
Gliricidia sepium	63	55	4.0	3.6	1.6
Psophocarpus palustris	54	20	5.1	3.0	0.9
Sesbania rostrata	47	ND[a]	4.8	3.7	ND
Leucaena leucocephala	45[b]	28[b]	3.4	5.0	2.6
LSD 0.05	12	18	1.6	0.6	0.5

a Not determined
b The low dry weight may have resulted from loss of leaves during handling

TABLE 5

Effect of application rate and method of application of dry leucaena leaves
and N on grain yield of maize in field trials

Treatment	Maize Yield (T ha^{-1})		
	1st Season (B$_1$)	2nd Season (B$_1$-residual)	Total
Leucaena ($\bar{x}$ of all leucaena treatments)	3.49	4.11	7.60
No N (control)	3.12	3.09	6.21
Difference	0.37	1.02*	1.39
Incorporation of dry leucaena leaves	3.43	4.09	7.52
Surface application of dry leucaena leaves	3.54	4.11	7.65
Difference	0.11	0.02	0.13
Dry leucaena leaves applied at planting	3.49	4.09	7.58
Dry leucaena leaves applied 1/3 at planting and 2/3 three weeks after planting	3.38	4.12	7.60
Difference	0.01	0.03	0.02
50 kg N ha^{-1} dry leucaena leaves	3.41	3.98	7.39
100 kg N ha^{-1} dry leucaena leaves	3.57	4.23	7.80
Difference	0.16	0.25	0.41
Leucaena ($\bar{x}$ of 50 and 100 kg N ha^{-1})	3.49	4.11	7.60
CAN[+] ($\bar{x}$ of 50 and 100 kg N ha^{-1})	3.66	3.23	6.89
Difference	0.17	0.88**	0.71

*, ** Significant at .05 and .01%, respectively
+ Calcium ammonium nitrate

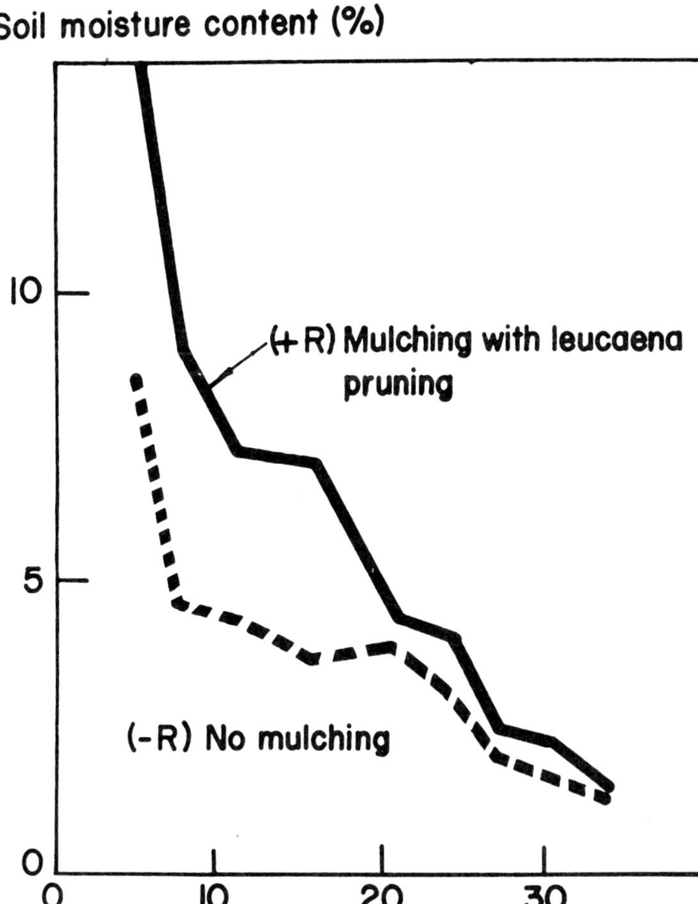

FIGURE 12   Soil moisture content of Apomu loamy sand (Psammentic Ustorthent) sampled at 5–15 cm depth with and without leucaena mulch.

In field trials Read (1982) could find no consistent trend between green manure and mulch in the first crop (Table 5) and attributed the results to high soil nitrogen levels. However, in the second season the crop grown without application of more nutrients showed that leucaena had a better residual effect than calcium ammonium nitrate (CAN).

On-going research supports the observation that nitrogen use efficiency is better when tree leaves are used as green manure rather than mulch, but such a generalization cannot be accepted. There are various situations in the tropics where the benefits from mulch outweigh those of green manure. On highly erodible soils under high rainfall conditions, mulching reduces soil erosion (Lal, 1974). Leucaena prunings applied as a mulch in alley cropping has resulted in better soil moisture retention (Fig. 12).

High soil temperatures have been found to inhibit germination and retard early growth of maize but when a mulch is used, soil temperatures were reduced and germination and early growth were improved (IITA, 1984).

Because both green manures and mulches are beneficial in most tropical areas, alley cropping techniques, in which species that are good green manure sources can be combined with species that are good mulch sources, are being developed.

CONCLUSIONS

Alley cropping brings to the tropics an effective means of using trees and woody shrubs to hasten biological soil fertility restoration. The leaves and twigs, through which nutrients are returned to the soil, can be applied as green manure or mulch. There are indications that nitrogen use efficiency is higher with green manuring than with mulching. However, there are situations in which mulching may be preferred to green manuring. Fortunately alley cropping is compatible with both green manuring and mulching.

Compared to herbacious legumes, which usually die leaving only dry tissues after the long dry period that precedes the major planting season in most tropical regions, the trees in alley cropping remain functional in the dry season providing fresh material for green manure or mulch. They also, to some extent, protect the environment from the fires that commonly destroy the dry residues of dead herbacious fallows.

Besides soil fertility restoration, alley cropping can be managed to prevent soil erosion and other forms of environmental decay and to provide firewood, a renewable and much needed energy source in the tropics.

References

Bene, J.C., Beal, H.W. & Cote, A. (1977). *Trees, Food and People. Land Management in the Tropics.* IDRC; Ottawa, Canada.

Benneh, G. (1972). Systems of agriculture in tropical Africa. *Economic Geography,* **48**, 244–257.

Combe, J., Jimenez Soa, H. & Claudia Monge (Comps) (1981). *Bibliography on Tropical Agroforestry.* CATIE; Turialba, Costa Rica.

De las Salas, G. (1979). Proceedings, Workshops: Agroforestry Systems in Latin America. CATIE; Turialba, Costa Rica.

Getahun, A., Wilson, G.F. & Kang, B.T. (1982). The role of trees in farming systems in the humid tropics. In *Agroforestry in the African Humid Tropics* (L.H. MacDonald, ed.), pp. 28–35. The United Nations University, Tokyo.

Guevarra, A.B., Whitney, A.S. & Thompson, J.R. (1978). Influence of intra-row spacing and cutting regimes on growth and yield of leucaena. *Agronomy Journal,* **70**, 1033–1037.

Huxley, P.A. ed. (1983). *Plant Resources in Agroforestry*. International Council for Research in Agroforestry; Nairobi; Kenya.

IITA (1982). Annual Report for 1981. Pp. 27–35. Ibadan, Nigeria.

IITA (1984). Effects of in situ mulch on soil temperature and maize production. Research Highlights for 1983, Ibadan Nigeria. Pp. 34–37.

Kang, B.T., Sipkens, L., Wilson, G.F. & Nangju, D. (1981a). *Leucena leucocephala* (Lam.) de Wit prunings as nitrogen source for maize (*Zea mays* L.). *Fertilizer Research*, **2**, 279–287.

Kang, B.T., Wilson, G.F. & Sipkens, L. (1981b). Alley cropping maize (*Zea mays*) and leucaena (*Leucaena leucocephala* Lam. de Wit) in Southern Nigeria. *Plant & Soil*, **63**, 165–179.

King, J. (1979). *Concepts of Agroforestry*. Proceedings of the International Conference on International Cooperation in Agroforestry. Nairobi, Kenya. July, 1979. International Council for Research in Agroforestry.

Kowal, J.M. & Kassam, A.K. (1978). *Agricultural Ecology of Savanna. A Study of West Africa*. Clarendon Press; Oxford, England.

Koudoro, D. (1982). Evaluation of four woody fallow species for alley cropping with maize and cowpeas. M.Sc. thesis, National University of Benin. Cotonou, Republic of Benin.

Lal, R. (1974). *The Role of Mulching Techniques in Tropical Soil and Water Management*. Technical Bulletin No. 1. IITA; Ibadan, Nigeria.

MacDonald, L.H. ed. (1982). *Agroforestry in the African Humid Forestry*. The United Nations University; Tokyo.

Milsum, J.N. & Bunting, B. (1928). Cover crops and manure. *Malayan Agricultural Jounral*, **26**, 256–283.

Mongi, H.O. & Huxley, P.A. (1979). *Soil Research in Agroforestry*. ICRAF; Nairobi, Kenya.

NAS (1977). *Leucaena: Promising Forage and Tree Crop for the Tropics*. National Academy of Science; Washington.

NAS (1979). *Tropical Legumes: Resources for the Future*. National Academy of Science; Washington.

Poulson, G. (1978). *Man and Tree in Tropical Africa*. IDRC-101e. IDRC; Ottawa, Canada.

Read, M. (1982). Management alternatives for maize-bean and Leucaena based cropping systems. Ph.D. thesis, Colorado State University; Fort Collins, Colorado.

Ruthenberg, H. (1971). *Farming Systems in the Tropics*. Oxford University Press; London.

Sanger, C. (1977). *Trees for People*. IDRC; Ottawa, Canada.

Sumberg, J.E. (1984). Small ruminant feed production in a farming systems context. Workshop on Small Ruminant Production Systems in the Humid Zone of West Africa, Ibadan, Nigeria, January 1984. International Livestock Centre for Africa.

Vine, H. (1953). Experiments on the maintenance of soil fertility at Ibadan, Nigeria. *Empire Journal of Experimental Agriculture*, **21**, 65–85.

Wassink, J.T. (1977). *Agroforestry: Interaction of Agriculture and Forestry for the Benefit of Man and his Environment*. Royal Tropical Institute, The Netherlands.

Webster, C. & Wilson, P.N. (1980). *Agriculture in the Tropics* (2nd edn.) Pp. 200–212. Longman, London.

Wilson, G.F. (1978a). Effect of an in-situ mulch on tomato production. First International Symposium on Tropical Tomato. October 1978, Shanhua, Taiwan, Republic of China. Asian Vegetable Research and Development Centre.

Wilson, G.F. (1978b). A new method of mulching vegetables with the in situ residue of a tropical cover crop legume (Abstract). Twentieth International Horticultural Congress, Sydney; Australia.

Wilson, G.F. & Kang, B.T. (1981). Developing stable and productive biological cropping systems for the humid tropics. In *Biological Husbandry. A Scientific Approach to Organic Farming* (B. Stonehouse ed.), pp. 193–203. Butterworth; London.

Nitrogen Fixation by Legumes and their Role as Sources of Nitrogen for Soil and Crop

J.N. Ladd, J.H.A. Butler and M. Amato

CSIRO, Division of Soils, Glen Osmond, South Australia, 5064, Australia

INTRODUCTION

In Australia wheat is grown in semi-arid areas characterized by low and variable rainfall, which frequently limits crop yields. Soils are old and often are initially nutrient deficient. In the southern wheatlands heavy reliance is placed on winter and spring rains, and phosphorus and trace element requirements are met by regular fertilizer applications. The maintenance of adequate levels of nitrogen for wheat growth has traditionally rested on symbiotic N_2 fixation by cultivars of the annual legumes *Medicago spp.* or *Trifolium subterraneum* grown in mixed pastures in rotations with the cereals. Long-term rotation trials have demonstrated improved yields of wheat grain of higher protein contents, and net gains in soil organic N levels (Greenland, 1971; Clarke & Russell, 1977).

Practices and conditions have changed in recent years. Increased frequency of cereal cropping, deterioration of legume pastures (fall in legume densities, attack by pests), harvesting of medic seed, and the growth and harvesting of grain legumes in the rotations are all seen as pressures contributing to the possible inadequacy of the symbiotic process to meet nitrogen demands. Consequently fertilizer N is now used more frequently to supplement N inputs due to symbiotic fixation.

There is a great need to quantify gross inputs and losses of nitrogen from all sources, and to determine the efficiency of N use by plants throughout the rotations. Some of the key processes of N cycling in soils (e.g. fixation, mineralization, nitrification, immobilization and dentrification) are determined by the activities of the soil biota. [15]N methodology is well suited to quantify the fate of N from specific sources (Hauck & Bremner, 1976). We have used ^{15}N-based techniques to help describe the role of legumes in field experiments relevant to the semi-arid, cereal-growing regions of southern Australia.

NITROGEN FIXATION BY LEGUMES

Integrated measures of the amounts of N fixed from the atmosphere or taken up from soil by legumes can be obtained by using ^{15}N isotope-dilution techniques (Hauck & Bremner, 1976; Rennie *et al.*, 1978; Knowles, 1981). By labelling the soil organic N pool with residues from applied fertilizer or plant materials it is possible to achieve conditions under which the atom % enrichment of mineralized N, even in soil profiles in the field, remains approximately constant over several months, thus meeting an important criterion for the valid use of the ^{15}N-dilution technique (for example, Ladd & Amato, 1985a).

Table 1 shows some results of pot experiments in which legume (*M. littoralis* var. Harbinger) and ryegrass (*Lolium multiflorum*) plants (4 per pot) were grown separately in ^{15}N-labelled soils with different N mineralization capacities, and in a soil supplemented with [15]NO_3^- of the same atom % enrichment as the [15]NO_3^- released from the labelled soil organic matter. For each soil the amount of N taken up by ryegrass, as determined by Kjeldahl digests of the whole plant, very nearly equalled that taken up by the legume, as determined from measurements of total legume N and the relative ^{15}N atom % enrichments of legume and ryegrass N. Ryegrass N accounted for usually more than 87% of the inorganic N of unplanted soils incubated under the same conditions as planted soils. The results suggest that the ryegrass and legume plants equally exploited the soils in the pots for inorganic N, and that recoveries of plant materials at harvest were nearly complete. The extra N of

TABLE 1

Inorganic N of unplanted soils, and N of a legume *M. littoralis* and of ryegrass
(*L. multiflorum*) grown separately in ^{15}N-labelled soils for 12 weeks

Soil	Inorganic –N of unplanted soil	N taken up from soil by rye grass	N taken up from soil by legume	N fixed by legume
Northfield	14.8	14.3	15.7	27.6
Caliph	23.1	20.0	18.9	54.5
Roseworthy 1	34.8	25.4	25.0	151.5
Avon 1	73.9	74.3	71.3	64.1
Avon 1 + NO_3^- (30)	104	97.1	97.5	98.3
Avon 1 + NO_3^- (90)	164	142.6	143.9	19.3

All values expressed as mg N per pot

the legume plants can be reasonably attributed to symbiotic N_2 fixation.

Studies by Butler & Ladd (1985a, b) on the fixation and uptake of N by legumes grown in several South Australian soils in pots may be summarized as follows.

(1) The percentage of N due to fixation in different parts of legume plants increased in the order roots < leaves < stems < pods.

(2) The amounts of N fixed per pot increased, and per plant decreased, with increasing legume numbers.

(3) The amounts of N fixed per pot, and per legume plant, decreased with increasing competition from ryegrass. The percentage decrease per plant for a given soil was directly related to the percentage increase in ryegrass dry matter.

(4) The amounts of N fixed per pot, and per legume plant, decreased with increasing concentrations of available nitrogen in soils.

(5) The amounts of fixed N were directly related to legume dry matter yields, irrespective of the soil, plant numbers and competition from ryegrass.

(6) The proportions of legume ± ryegrass tops required to be returned to soil to effect a net N gain increased with increasing concentrations of available N in the soil.

"Pot" studies have been extended to the field in experiments in which replicated samples (30 kg dry weight) of ^{15}N-labelled soils were placed in large steel or plastic cylinders (300 mm diam. × 400 mm), open-ended and free-draining but fitted with a woven polyester fabric to act as a root barrier at 300 mm depth. The cylinders were installed in soils to 350 mm depth. The labelled soils were oversown with legume and ryegrass seed as appropriate, and seedlings were later thinned to the required numbers. At sowing the soils also received a rhizobial inoculum and an amendment of superphosphate and trace elements.

Such an experiment was conducted at Avon, South Australia, using two topsoils (Roseworthy 2 and Avon 2) containing organic ^{15}N, formed by immobilizing $^{15}NH_4^+$ by incubation with sucrose for 9 weeks at 25°C. Properties of each soil are shown in Table 2. Both soils were calcareous and of similar pH and texture. The concentrations of organic C and N of Avon 2 soil exceeded those of Roseworthy 2, but the differences were not as great as in earlier batches of soils from the same area (Butler & Ladd, 1985a). In both soils, the ^{15}N atom % enrichments of soil inorganic N at sowing exceeded those at crop harvest (Table 2), indicating that the labelled and unlabelled organic sources of available N were less equilibrated than preferred for a precise application of the ^{15}N-dilution technique. Triplicated cylinders of each soil were sown with either 30 *M. littoralis* plants, to be grown alone, or together with 30 ryegrass plants. After 24 weeks under field conditions plants were

TABLE 2

Properties of two soils used in a field experiment to measure N_2
fixation by the legume, *Medicago littoralis*

Soil property	Roseworthy 2	Avon 2
Organic C (%)	1.4	1.8
Calcium carbonate (%)	2.7	10.3
pH	8.3	8.5
Total N (%)	0.106	0.129
^{15}N atom % excess of Total N	0.160	0.202
^{15}N atom % excess of Inorganic N (at sowing)	0.680	1.235
^{15}N atom % excess of Inorganic N (at harvest)	0.343	1.003
$NaHCO_3$-extractable P (μg g^{-1})	15	35
Clay (%)	14	22
Silt (%)	3	6
Sand (%)	80	59

TABLE 3

Fixation and uptake of nitrogen by a legume* and a legume-grass*
mixture grown in two soils at the same field location

Soil	Plant	Plant part	Plant-dry matter (t ha^{-1}) Sub-Total	Total	Plant N (kg ha^{-1}) fixed from atmosphere Sub-Total	Total	taken up from soil Sub-Total	Total
Roseworthy 2	legume	T**	12.51		148.0		98.3	
		R	0.79	13.30	4.2	152.2	10.3	108.6
	legume	T	2.74		43.7		11.6	
		R	0.18	2.92	1.3	45.0	1.1	12.7
	+			15.25		45.0		118.8
	grass	T	10.78		0.0		87.8	
		R	1.56	12.33	0.0	0.0	18.3	106.1
Avon 2	legume	T	10.28		87.5		121.1	
		R	0.82	11.10	7.6	95.1	7.8	128.9
	legume	T	0.86		20.6		3.4	
		R	0.13	0.98	1.7	22.3	0.9	4.3
	+			19.87		22.3		207.4
	grass	T	16.90		0.0		182.8	
		R	1.99	18.89	0.0	0.0	20.3	203.1

*Legume, *Medicago littoralis*; Grass, *Lolium multiflorum*
**T=tops, R=roots

harvested, the roots being recovered by washing and sieving of the soils which had been dug out of the cylinders to the original incorporation depth (300 mm).

Table 3 summarises the plant recovery data. Yields of plant dry matter and N were considerably higher (2–3 times) than those frequently obtained under field conditions, but this was not unexpected given the season and the greater depth (30 cm) of topsoil potentially available in the cylinders for root growth.

The ^{15}N atom % enrichments of ryegrass tops N were slightly (6–10%) higher than those of ryegrass roots N and, as anticipated, were intermediate between those of soil NO_3^-N sampled at sowing and harvest. The enrichment of tops ryegrass N was used as the reference for calculating N uptake by legumes.

For the legume grown alone in Roseworthy 2 soil (of the lower organic matter content), the amounts of N_2 fixed were greater and the amounts of N taken up were less than in Avon 2 soil. Ryegrass growth depressed legume dry matter yields, the amounts of N_2 fixed, and especially the amounts of N taken up from soil by the legume. These effects were greater in the Avon 2 soil where conditions were more suited to ryegrass growth. Thus in the Roseworthy soil ryegrass decreased the amounts of N_2 fixed by 70%, and the proportions of legume N due to fixation increased from 58% to 78%. Corresponding values for the Avon soil were 77%, and from 42% to 84%.

The field study results are consistent with those previously reported for greenhouse studies but differed, in the case of the Avon soil, in the following respect. In the field study the total amount of soil-derived N taken up from the Avon soil by the legume-ryegrass mixture (207 kg ha^{-1}) was considerably more than that (129 kg ha^{-1}) taken up by the legume grown alone. In greenhouse studies, irrespective of the combinations of legume and ryegrass grown alone or together in pots, the total N taken up from soils was similar (Butler & Ladd, 1985a). Clearly in the field where conditions would permit leaching of NO_3^- perhaps to depths below the root barrier in the cylinders, the ryegrass in the Avon soil was better able than legume to utilize potentially-available N. Thus differences between the total N taken up by legume alone and by legume plus ryegrass) 79 kg ha^{-1}) were only partly explained by differences in residual NO_3^- of the respective topsoils. At harvest of the legume-grass mixture, the Avon soil contained 14 kg NO_3^-N ha^{-1}, i.e. 24 kg NO_3^-ha^{-1} less than in the Avon soil in which had grown legume alone. This deficit accounted for only 31% of the difference in N uptake by plants.

In the case of the Roseworthy soil, plant N uptakes were similar for both treatments (109 and 119 kg ha^{-1}), as were concentrations of soil NO_3^-N at harvest (13 and 8 kg ha^{-1}).

The proportions of plant tops N required to be returned to soil to effect a net gain in soil N ranged widely in the four treatments. In increasing order they were legume alone in Roseworthy soil (38%), legume alone in Avon soil

(54%), legume plus ryegrass in Roseworthy soil (68%), and legume plus ryegrass in Avon soil (89%).

DECOMPOSITION OF LEGUME RESIDUES

Field practices lead to variable returns of pasture residues. In some areas grazed green pastures may be incorporated in spring allowing about 9–10 months fallow before sowing wheat in the following year. More commonly the pastures under variable grazing pressures are allowed to senesce and weather on the soil surface. The soil is cultivated in the following autumn, to allow 2–3 months of fallow before sowing.

(a) Surface litter

Table 4 shows the weights and some properties of residual surface litter following senescence and weathering of ungrazed field plots of the legume *M. scutellata* grown alone, or grown as a mixed sward with ryegrass and barley grass (*Hordeum* sp.). These studies were part of a larger experiment (see Page 282) to determine the effects of rotations on the amounts and distribution of NO_3^- in soil profiles.

In late spring (October) when the plant litter decomposition measurements commenced, the vegetation was green and the legumes were flowering and with pods set; but there were indications of incipient moisture stress. Six weeks later in early summer the surface litter was brown; recovered plant C and dry matter had decreased substantially (20–30%), and plant N even more or so (40%) (Table 4). During this browning-off period, the percentage N content of plant tops decreased (legume alone, from 2.04 to 1.55%, legume-grass mixture, from 1.49 to 1.23%), and C:N ratios increased (legume alone, from 21.1 to 26.9:1; legume-grass mixture, from 29.8 to 35.9:1). Thereafter, for a further 17 weeks until cultivation in the following year, changes in the amounts of recovered plant dry matter, C and N were relatively small. Occasional rain during summer and early autumn permitted brief periods of plant growth. Nevertheless the amounts of N remaining in the dry legume residues at cultivation averaged $72\,kg\,N\,ha^{-1}$, i.e. about 50% of that present at the commencement of the net decomposition phase. The corresponding amounts present in the dry legume-grass litter averaged $63\,kg\,N\,ha^{-1}$, about 70% of the initial amounts.

Both the green legume tops and the senesced brown litter contained hot water-soluble organic compounds of an average C:N ratio of 10–11:1. The proportions of legume C and N extractable by hot water decreased markedly during the initial browning-off period (Table 4). Thus the net amounts of

TABLE 4

Weights and properties of plant tops during senescence and weathering of a legume
and a legume-grass pasture under field* conditions

Pasture	Period after late flowering (weeks)	Weight of plant tops (kg ha^{-1})			Properties of plant tops		
		Dry matter	C	N	C:N ratio	Extractable C** (% of plant C)	Extractable N** (% of plant N)
Legume (*M. scutellata*) alone	0	6820	2950	140	21.1	19.5	37.5
	6	5470	2280	85	26.9	8.9	21.7
	12	4040	1680	59	28.6	6.9	19.5
	23	4500	1880	72	26.2	8.8	20.9
Legume-ryegrass-barley grass mixture	0	6170	2710	92	29.8	–	–
	6	4290	1890	53	35.9	–	–
	12	4760	2070	51	40.3	–	–
	23	5060	2090	63	33.6	–	–

*pastures not grazed
**C and N extractable with water at 60°C for 30 min.

TABLE 5

Decomposition under field conditions of ^{15}N-labelled legume (*M. littoralis*)
tops applied to the soil surface

Plant material	Decomposition time (weeks)	Plant dry matter recovered from surface	Legume N recovered		
			from surface material	as organic N in soil	as inorganic N in soil
Legume	4	98.8	74.9	15.7	3.2
	8	95.6	66.5	23.4	4.2
	19	96.7	57.7	52.5	5.2
Legume-grass mixture	8	83.0	70.9	23.4	4.2
	19	77.5	62.0	18.6	3.3

All recoveries expressed as percentages in input
*Mixture of legume and barley grass in ratio 1:2, w/w: replicates lost for 4 week sampling

soluble C and N which disappeared were substantial, and were equivalent to 55% and 62% respectively of the total legume C and N which disappeared from the litter during this period. The extent to which the losses of plant litter C and N can be attributed to leaching of soluble organic components, or to fragmentation and/or decomposition of the litter, is unknown. The C:N ratio of the materials which disappeared from the legume litter during the first six weeks averaged 12.2:1, suggesting that if any decomposition accompanied by microbial growth had occurred, then this would have led to rapid net mineralization of some of the legume tops N. This process together with any action of plant deaminases may have provided opportunities for losses of N by volatilization also.

In a separate experiment ^{15}N-labelled legume (*M. littoralis*) tops were applied to the surface of soils and allowed to decompose under field conditions during the mainly dry months of summer and early autumn. Table 5 shows the recoveries of ^{15}N in litter and topsoil. Most of the applied litter ^{15}N was recovered in plant plus topsoil, even after exposure in the field for 19 weeks. Only small amounts of the ^{15}N in topsoils were present as inorganic ^{15}N. The recovery of legume ^{15}N in dried litter was marginally affected by the presence of grass residues. The results suggest that decomposition of the legume litter had been slight, and that the accession of ^{15}N to topsoils may well have been due to labelled plant fragments remaining on the soil surface after handpicking.

The experimental procedures with labelled legume tops differed in several, perhaps important, respects from those with unlabelled legume plots discussed above, and the results are not directly comparable. The experiments were run at the same field site and in the corresponding seasons but in different years, resulting in different climatic (especially rainfall) patterns (Table 6).

TABLE 6

Mean air temperature and rainfall for periods of decomposition of
plant litter in two field experiments

Plant litter	Decomposition period (weeks)	Rainfall (mm)	Mean air temperature (°C)
Legume and	0–6	48.0	19.0
Legume-grass	6–12	26.9	24.6
Mixture	12–23	25.5	22.7
(Unlabelled)			
Legume and	0–4	39.3	20.4
Legume-grass	4–8	3.8	22.8
Mixture	8–19	12.8	22.4
(^{15}N-labelled)			

The labelled legume tops did not dry under field conditions but were grown to the pod development stage in the greenhouse, then oven-dried before applying to the soils in the field. Further, these soils were contained within open-ended plastic cylinders inserted to 15 cm depth, and projecting 2.5 cm above the soil surface. The tops of the cylinders were covered with nylon netting (1 cm^2 mesh) to protect the applied plant material from wind-blow. Despite these differences in techniques it seems likely that at least part of the disappearance of legume litter C and N accompanying senescence in the larger-scale plots may have been due to the failure to recover fallen and perhaps fragmented legume top components of lower-than-average C:N ratios.

(b) Incorporated residues

The central role of the soil fauna and microflora in decomposing plant residues in soils has long been established. Thus the net rates of decomposition will be determined by factors which influence the availability to biological attack of substrate and product, and which influence the activities and turnover of the decomposer populations. The patterns of decomposition of legume materials mixed and incubated with soils are typical of those determined for other plant materials. Characteristically there is initially a rapid decline in the concentrations of organic residues followed by progressively slower rates of decay of the more resistant plant compounds and microbial metabolic products (Jenkinson, 1981).

The extent of decomposition during the initial phase of rapid decay varies with the legume plant part, and is directly related to the proportions of plant C extractable with hot water (Amato *et al.*, 1984). Similarly green legume tops with higher proportions of water-soluble C (and N) underwent a more extensive decomposition initially when incubated with moist soil than did the brown, senesced and weathered legume litter, although differences became less marked with time. Specifically, after incubating ground legume top material in a moist soil for 7 days at 25°C, 35% of green legume top C and 18–24% of brown legume litter C was evolved as CO_2. After 84 days incubation, 65–70% of the C of the green and brown legume materials had been converted to CO_2.

Decomposition of the legume materials in incubated moist soils was more extensive than the disappearance of legume litter under summer field conditions, and probably involved substrates of high average C:N ratios, with marked effects on the release of inorganic N. Figure 1 shows that for a soil amended with green legume tops and incubated, there is a net release of inorganic N throughout. But the percentage release, based on the difference between the release in the amended soil and an unamended soil control was only 14% of plant N input after six weeks, compared with a 40% disappearance of green legume tops N after exposure on the soil surface for six weeks in the field. Further, when brown legume litter materials containing smaller amount of easily-decomposable, water-soluble components of low

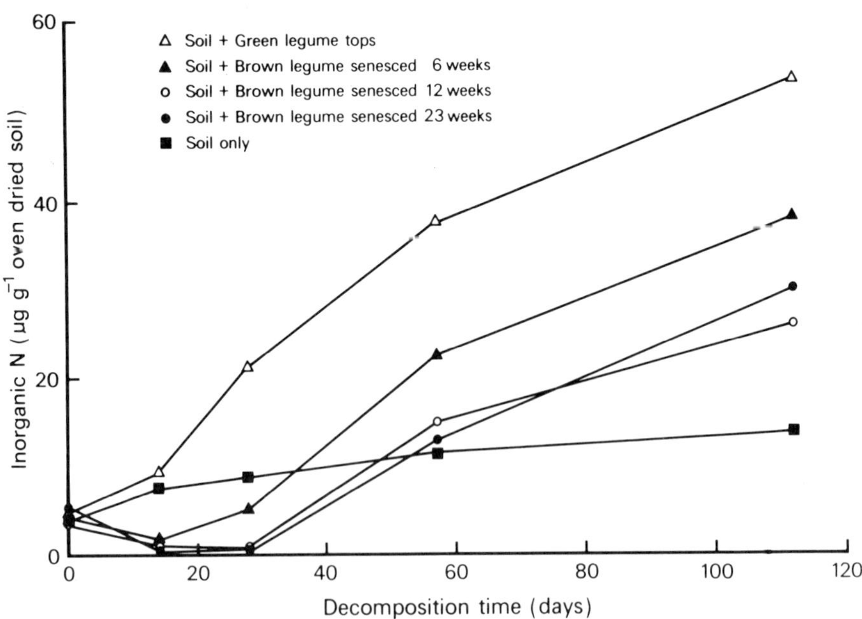

FIGURE 1 Mineralization of nitrogen from a soil amended with green and senesced legume tops.

C:N ratios were incubated in soils, their decomposition caused a temporary net immobilization of N. Incubation of that legume litter which previously had senesced and weathered for 23 weeks in the field (and which at cultivation had been incorporated in topsoils—see page 282), led to net N mineralization only after 7–8 weeks.

The use of [14]C- and ^{15}N-labelled plant materials enables assessments to be made of the average net rates of decay of organic residues formed long after the initial substrates have been metabolized. Figure 2, adapted from Ladd *et al.* (1981a, 1985) shows the percentage retention of ^{14}C and ^{15}N in organic residues and in the soil biota, following the decomposition over eight years of a mixture of tops and roots of *M. littoralis* incorporated at a field site (Roseworthy) in South Australia.

The pattern of decomposition showed the characteristic rapid initial decline, followed by a period of slow decrease. About 30% of legume [14]C

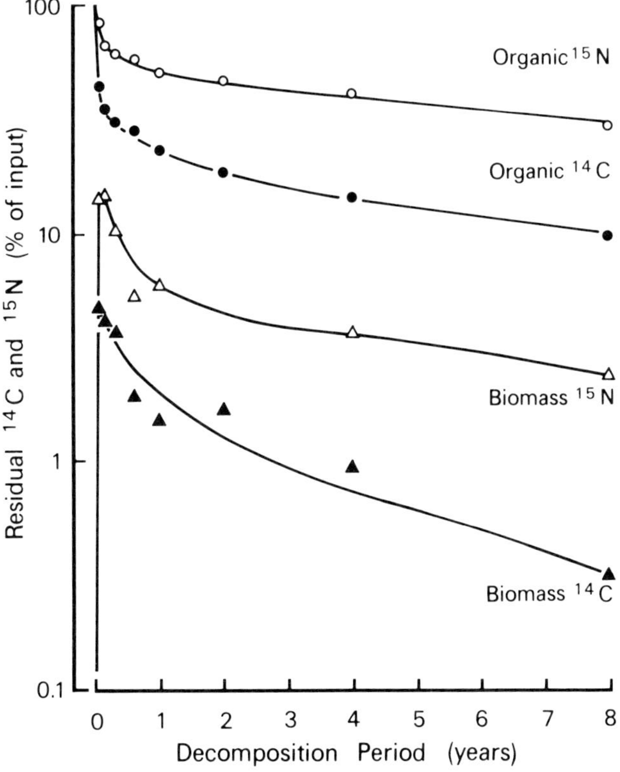

FIGURE 2 Formation and decline of isotope-labelled biomass C and N during the decomposition of ^{14}C, ^{15}N-labelled legume material.

remained in the soil after 0.3 years, declining to about 11% of ^{14}C input after eight years. The average rate of decomposition of legume residues at several South Australian sites was approximately double that averaged for ryegrass decomposition at some English sites, and about one half the rate established for decomposition under Nigerian conditions (Jenkinson & Ayanaba, 1977; Ladd & Amato, 1985a; Ladd et al., 1985). Temperature may have been the dominating influence on decay rates.

Differences between the soils in the percentage retention of organic residues generally are small and may be due in part to factors which influence biomass C and N turnover, e.g. protection or stabilization of the decomposer populations (van Veen et al., 1984).

Figure 2 also shows that the patterns of decay of organic ^{15}N residues and of changes of biomass ^{15}N are similar to those respectively of ^{14}C-labelled residues and biomass ^{14}C. After 0.3 years about two thirds of legume ^{15}N remained in soils as organic residues, declining to about one third after eight years. The C:N ratio of the added legume material was 15.3:1 and it was anticipated that net mineralization of legume N would occur concurrently with plant decomposition. Since the nitrogen of added plant materials appears to enter the immobilization-mineralization turnover cycle rapidly (Bartholomew, 1965), the retention in organic residues of applied ^{15}N will be influenced by (1) the relative amounts of available C and N from both added substrate and from soil organic matter, and (2) the degree of equilibration of the N released from both sources. Thus under the conditions of our experiment, as reported in Figure 2 (Ladd et al., 1985), about 40% of the legume N at a maximum could have been available for crop use in the year following the return of legume material to soil; and only a further 25% made available over the next seven years. Legume materials of higher C:N ratios (e.g. weathered surface litter) could be anticipated to release lower proportions of their nitrogen for uptake by subsequent crops. Because N released from plant residues and soil organic matter sources may at least partly equilibrate before entering immobilization reactions, it is possible for ^{15}N of plant materials to be released as inorganic ^{15}N to some extent at a time when net immobilization of soil-derived N is occurring.

AVAILABILITY TO WHEAT CROPS OF NITROGEN RELEASED FROM DECOMPOSING LEGUME RESIDUES

The supply of N from decomposing legume residues to successive wheat crops, and the total recovery of residue N in soil plus crop have been measured at several field sites in South Australia by using ^{15}N-labelled M. littoralis as source material (Ladd et al., 1981b, 1983; Ladd & Amato, 1985b). Mixtures of the legume tops and roots were mixed with topsoils confined within open-ended steel cylinders which were installed to 90 cm depth in undisturbed soils. The C:N ratios of the legume mixtures ranged from 11.1 to 14.9:1; and plant N inputs ranged from 25 to 100 kg ha^{-1}, which are realistic levels for most field conditions in the southern Australian wheatlands. The added legume materials were allowed to decompose for 5–8 months before sowing wheat. Where a second crop was grown, soils contained decomposing roots of the previous crop as well as other ^{15}N-labelled residues. Results may be summarized as follows.

(1) The percentage recovery of legume-derived ^{15}N in wheat crops plus soil profiles to 90 cm depth generally exceeded 90%. Leaching was probably an important factor in experiments where lower ^{15}N recoveries were obtained, although this is generally not considered an important process for N loss under Mediterranean-type climatic conditions (Noy-Meir & Harpaz, 1977).

(2) Residual ^{15}N in cropped soils was mainly in organic forms.

(3) By the flowering stage of wheat growth residual organic ^{15}N concentrations in cropped soils (most of the plant roots having been removed), exceeded those in unplanted soils. This is believed to be due either directly to the return of crop residues or to a greater immobilization of inorganic ^{15}N. The latter would reflect an enhanced microbial metabolism accompanying decomposition and turnover of root C (Martin & Puckridge, 1982).

(4) The percentage uptakes of legume-derived ^{15}N by first and second wheat crops were approximately independent of the amounts of legume N applied (in the range 25–100 kg N ha^{-1}).

(5) The percentage uptake of ^{15}N by a first wheat crop at tillering was about 75% that at flowering. There was no net loss of ^{15}N from wheat tops during ripening.

(6) The percentage uptake of legume-derived ^{15}N by first wheat crops ranged widely (10.9–27.8%) for different sites in different seasons and were related to grain yields in soils of similar profile properties.

(7) In a given season, the percentage uptake of legume-derived ^{15}N to a first wheat crop (17%) was about one third that of fertilizer ^{15}N; the availabilities of ^{15}N of legume and fertilizer residues to a second wheat crop declined but remained significantly different.

(8) Irrespective of wheat yields at different sites, N released from decomposing legume residues contributed only minor percentages of the available N pool; the contributions were approximately proportional to legume N input. In our experiments using legume materials of low C:N ratio (11.1–14.9:1) incorporated into topsoils of about 0.1% total N, 10% of the wheat N was derived from legume materials for each 50 kg ha^{-1} legume N input; the relative availability of legume N is halved for the succeeding crop (Ladd & Amato, 1985a).

(9) The addition of legume material to soils may significantly increase the mineralization of soil-derived organic N, and under appropriate conditions, the amounts of soil-derived N taken up subsequently by wheat crops.

MINERALIZATION OF NITROGEN AND ITS AVAILABILITY TO WHEAT IN ROTATIONS

Wheat crops following a legume pasture frequently yield more than crops following wheat, and yields may correlate with the amounts of inorganic N in soil profiles in sowing. Data in Table 7 from unpublished field experiments by Ladd *et al.* are consistent with such observations. In the year following a stand of *M. scutellata*, wheat crops yielded $2.8-2.9\times10^3$ kg ha^{-1} of grain dry matter and 85–93 kg ha^{-1} of tops N; the amounts of NO$_3^-$N in soil profiles to 60 cm at sowing were 108–111 kg ha^{-1}. By contrast, following a preliminary wheat crop, wheat crops yielded $1.8-2.2\times10^3$ kg ha^{-1} of grain dry matter and 59–82 kg ha^{-1} of tops N, and the amounts of NO$_3^-$N in soils at sowing ranged from 62 to 66 kg ha^{-1}. Wheat yields following a legume-grass pasture were $2.3-2.5\times10^3$ kg ha^{-1} of grain dry matter.

In the first year of the experiment in plots sown with the legume only, soil NO$_3^-$N levels fell (net) by 28 kg ha^{-1} between sowing and flowering, the latter time being shortly before senescence and browning of litter with the approach of summer. In the same period in corresponding plots of the mixed legume-

TABLE 7

Mineralization, and availability to wheat, of nitrogen following the growth and decomposition of residues from a wheat crop and from a legume and mixed legume-grass sward

Plant Cover year 1	Inorganic N (0–60 cm, kg ha^{-1}) in soil at					N in Wheat Tops (kg ha^{-1}) at Maturity year 2	
	Sowing year 1	Flowering year 1	Cultivation year 2	Sowing year 2			
Wheat	75	13	36	*a	66	a	72
				*b	62	b	59
Legume-grass	65	17	37	a	60	a	70
				b	64	b	58
Legume	69	41	71	a	111	a	93
				b	108	b	85

*a. Surface residues removed, or b. surface residues returned at cultivation. N (kg ha^{-1}) in surface residues at cultivation: 27 (wheat), 63 (legume-grass), 71 (legume)

grass pasture and of wheat, NO_3^-N concentrations fell by 48 and 62 kg ha^{-1} respectively. Thus at or near the end of N uptake by all plants, NO_3^-N levels in soils under the legume alone were greater by 24 and 28 kg ha^{-1} than in soils under legume-grass pastures and wheat respectively.

Thereafter until the time of sowing in the second year the differences in the amounts of NO_3^-N increased by 23 and 17 kg ha^{-1} respectively. The greater gains in the plots sown with legume alone during this period, and especially subsequent to cultivation, may have been partly due to the decomposition of legume top materials of low C:N ratio which fell and fragmented following 'browning-off' of the litter (see section (a), page 276).

No significant additional effects on soil NO_3^- or on wheat yields were obtained by returning recoverable wheat stubble or legume pasture litter at cultivation. In the case of legume litter, incubation studies under favourable conditions demonstrated that C and N metabolism led to an initial net immobilization of N for about two months before the onset of net mineralization (Fig. 1). Under field conditions decomposition may have been slower such that the major contribution of the legume litter N to the total available N pool may not have been expressed in the first crop after litter incorporation.

The data from the field plots have demonstrated a significantly ($P < 0.01$) greater accumulation of NO_3^-N in soil profiles following a legume pasture than following a mixed legume-grass pasture or a wheat crop (Table 7). The differences in soil NO_3^-N concentrations at the time of sowing in the year following the growth of legume alone or of wheat, were due more to the greater residual NO_3^-N in the legume-sown soils at the end of N uptake by plants in the first year, and less to the subsequent greater release of N in the soils containing legume residues.

The results are consistent with isotope data from field experiments. In the year following incorporation of green legume materials in topsoils, wheat crops take up $<30\%$ (and usually about 20%) of the nitrogen of the decomposing legume residues. Most of the legume-derived N remains in soil in relatively stable organic forms. From these consistent observations we have concluded that the primary value of legumes in the nitrogen economy of legume-cereal rotations is to maintain or increase the concentrations of organic N in soils to ensure an adequate delivery of nitrogen for cereal growth.

Of course the relative contributions to the available N pool from recently-added plant residues and from soil sources will be influenced by their concentrations and properties. For a range of legume N inputs normal for South Australian conditions, and for soil of commonly-found organic N levels, our studies demonstrate that the percentage contribution to the available N pool from the legume residues is small (10% per 50 kg ha^{-1} input of legume N), even within the first year of legume decomposition. In these studies the C:N ratio of the applied ^{15}N-labelled legume material was $<15:1$.

However, under normal field conditions the C:N ratio of legume residues would be expected to range widely. For example senesced legume residues, weathered on the soil surface over summer prior to cultivation in the wheat growing year, may have less readily decomposable components and be of higher C:N ratio than green legume materials, and consequently may contribute even less nitrogen for crop use.

By contrast in grazed pastures some legume N will be returned in faeces, and as urea N in animal urine, and any urea N not lost as NH_3 by volatilization would have an availability approaching that of fertilizer N.

ACKNOWLEDGEMENTS

The authors thank Mrs M.B. Bohnsack and Messrs P. Williams, G. Levendis and B. Swaby for technical assistance. Studies on nitrogen fixation are supported in part by the Australian Wheat Industry Research Council.

SUMMARY

The important Australian wheat-growing industry is based mainly in southern regions of the country where traditionally the maintenance of N supplies for cereal growth has depended upon symbiotic N_2 fixation by legumes, grown in rotations. Nitrogen inputs, transformations and uptake by crops are affected markedly by seasonal climatic conditions, with their strong influence on plant growth and biological activities.

The role of the legume as a supplier of N to soils and crops has been investigated using ^{15}N-labelled plant residues and soil organic matter. Quantitative data are presented on five aspects of the rotations:

(1) The use of a ^{15}N-dilution technique has been used in field and green-house studies to proportion legume N into symbiotically-fixed N and soil-derived N, thus permitting calculations of net gains to or losses from the soil of N from plant residues, as affected by agricultural practice.

(2) Measurements have been made of the disappearance during pasture senescence and weathering of surface litter C and N. Studies of the decomposition of ^{14}C, ^{15}N-labelled legume residues incorporated in topsoils, and of the formation and decline of labelled decomposer populations, have permitted estimates in long-term field experiments of organic matter retention and N release.

(3) Measurements of legume-derived N in successive wheat crops have permitted assessments of factors influencing the relative availability of N from fresh and older plant residues and from soil organic matter N.

(4) Measurements of total ^{15}N recovery at different growth stages, and in

the non-growth seasons, have permitted estimates of N losses in planted and unplanted situations.

(5) Measurements of inorganic N in soil profiles throughout rotations have demonstrated that the availability of N to cereal crops may be significantly influenced by the fate of NO_3^- N during the growth phase in previous seasons.

References

Amato, M., Jackson, R.B., Butler, J.H.A. & Ladd, J.N. (1984). Decomposition of plant material in Australian soils. II. Residual organic ^{14}C and ^{15}N from legume plant parts decomposing under field and laboratory conditions. *Australian Journal of Soil Research,* **22**, 331–341.

Bartholomew, W.V. (1965). In *Soil Nitrogen* (W.V. Bartholomew & F.E. Clark, eds.), pp. 285–306. American Society of Agronomy; Madison, Wisconsin.

Butler, J.H.A. & Ladd, J.N. (1985a). Fixed and soil-derived nitrogen in legumes grown in pots in soils of different available nitrate. *Soil Biology & Biochemistry,* **17**, 47–55.

Butler, J.H.A. & Ladd, J.N. (1985). Growth and nitrogen fixation by *Medicago littoralis* in pot experiments: Effect of plant density and competition from *Lolium multiflorum*. *Soil Biology & Biochemistry,* **17**, 355–361.

Clarke, A.L. & Russell, J.S. (1977). Crop sequential practices. In *Soil Factors in Crop Production in a Semi-Arid Environment* (J.S. Russell & E.L. Greacen, eds.), pp. 279–300. Australian Society of Soil Science Monograph. University of Queensland Press; St. Lucia.

Greenland, D.J. (1971). Changes in the nitrogen status and physical conditions of soils under pastures, with special reference to the maintenance of the fertility of Australian soils used for growing wheat. *Soils & Fertilizers,* **34**, 237–251.

Hauck, R.D. & Bremner, J.M. (1976). Use of tracers for soil and fertilizer nitrogen research. *Advances in Agronomy,* **28**, 219–226.

Jenkinson, D.S. (1981). The fate of plant and animal residues in soil. In *The Chemistry of Soil Processes* (D.J. Greenland & M.H.B. Hayes, eds.), pp. 505–561. John Wiley and Sons Inc.; New York.

Jenkinson, D.S. & Ayanaba, A. (1977). Decomposition of carbon-14 labelled plant material under tropical conditions. *Soil Science Society of America Journal,* **41**, 912–915.

Knowles, R. (1981). The measurement of nitrogen fixation. In *Current Perspectives in Nitrogen Fixation. Proceedings of the 4th International Symposium on Nitrogen Fixation* (A.H. Gibson & C.E. Kettering, eds.), pp. 327–333. Australian Academy of Sciences; Canberra.

Ladd, J.N. & Amato, M. (1985a). Nitrogen cycling in legume-cereal rotations. In *Proceedings of the International Symposium on Nitrogen Management in Farming Systems in the Humid Tropics,* Ibadan, Nigeria. In press.

Ladd, J.N. & Amato, M. (1985b). The fate of nitrogen from legume and fertilizer sources in soils successively cropped with wheat under field conditions. *Soil Biology & Biochemisty,* submitted.

Ladd, J.N., Oades, J.M. & Amato, M. (1981a). Microbial biomass formed from ^{14}C, ^{15}N-labelled plant material decomposing in soils in the field. *Soil Biology & Biochemistry,* **13**, 119–126.

Ladd, J.N., Oades, J.M. & Amato, M. (1981b). Distribution and recovery of nitrogen from legume residues decomposing in soils sown to wheat in the field. *Soil Biology and Biochemistry,* **13**, 251–256.

Ladd, J.N., Amato, M. & Oades, J.M. (1985). Decomposition of plant material in Australian soils. III. Residual organic and microbial biomass C and N from isotope-labelled plant material and soil organic matter decomposing under field conditions. *Australian Journal of Soil Research,* **23**, in press.

Ladd, J.N., Amato, M., Jackson, R.B. & Butler, J.H.A. (1983). Utilization by wheat crops of nitrogen from legume residues decomposing in soils in the field. *Soil Biology & Biochemistry,* **15**, 231–238.

Martin, J.K. & Puckridge, D.W. (1982). Carbon flow through the rhizosphere of wheat crops in South Australia. In *The Cycling of Carbon, Nitrogen, Sulfur and Phosphorus in Terrestrial and Aquatic Ecosystems* (I.E. Galbally & J.R. Freney, eds.), pp. 77–82. Australian Academy of Sciences; Canberra.

Noy-Meir, I. & Harpaz, Y. (1977). Agroecosystems in Israel. In *Agroecosystems; Special Issue, Cycling of Mineral Nutrients in Agricultural Ecosystems. Agroecosystems,* **4**, 143–167.

Rennie, R.J., Rennie, D.A. & Fried, M. (1978). Concepts of ^{15}N usage in dinitrogen fixation studies. In *Isotopes in Biological Dinitrogen Fixation*, pp. 107–133. IAEA: Vienna.

van Veen, J.A., Ladd, J.N. & Frissel, M.J. (1984). Modelling C and N turnover through the microbial biomass in soil. *Plant & Soil,* **76**, 257–274.

The Microbiological Requirements to Grow White Clover (*Trifolium repens*) in Hill Soils in Scotland

P. Newbould, Anne Rangeley and J.R. Jebb

Hill Farming Research Organisation, Bush Estate, Penicuik, Midlothian, EH26 0PY, Scotland

INTRODUCTION

The native hill and upland pastures of Scotland produce relatively little dry matter each year and much of it is only of moderate nutritional quality. These pastures support low populations of grazing ruminants, usually sheep, though cattle, red deer and goats are also found in some places. Hill sheep farming in these regions is sustained by subsidies provided under the Less Favoured Areas directive of the EEC. The economic viability of hill sheep farming can be enhanced by increased production of saleable animals following pasture improvement and the strategic use of the better quality herbage so as to best benefit the animal (Maxwell *et al*, 1979). Pasture improvement of this type is very dependent on the establishment, growth and fixation of nitrogen by white clover (*Trifolium repens*).

White clover does not grow in most Scottish hill soils without amelioration of their acidity and low levels of available nutrients. Thus, lime is required to raise the soil pH, and of the major nutrients, P and K can produce marked growth responses (Rangeley, 1980). Inoculation with the microbial symbionts, *Rhizobium trifolii*, to ensure effective nitrogen fixation, and vesicular arbuscular mycorrhizal fungi (VAM) to provide the plant with P, has been attempted but responses have been unpredictable.

This paper describes the Hill Farming Research Organisation's (HFRO) experience in the study of responses of white clover to inoculation with rhizobia and VAM, and interactions between the symbionts and the host plant.

STUDIES AT HFRO

Lime and Nutrients

Initially work was directed at establishing the soil pH and nutrient requirements to grow white clover which fixed N_2 on hill soils. It was found that dressings of between 2.5 and 7.5 t lime ha^{-1} were needed to raise the pH above 5.2 and, ideally, into the range 5.5–6.0. Responses to additions of phosphorus and potassium were found to be linked as shown by the data in Figure 1. For a deep peat soil from Lephinmore, Argyll, there was little response in growth of white clover to applications of phosphorus fertiliser in the absence of potassium. As the quantity of the latter element was increased so did the response to phosphate. For economic reasons the quantities of

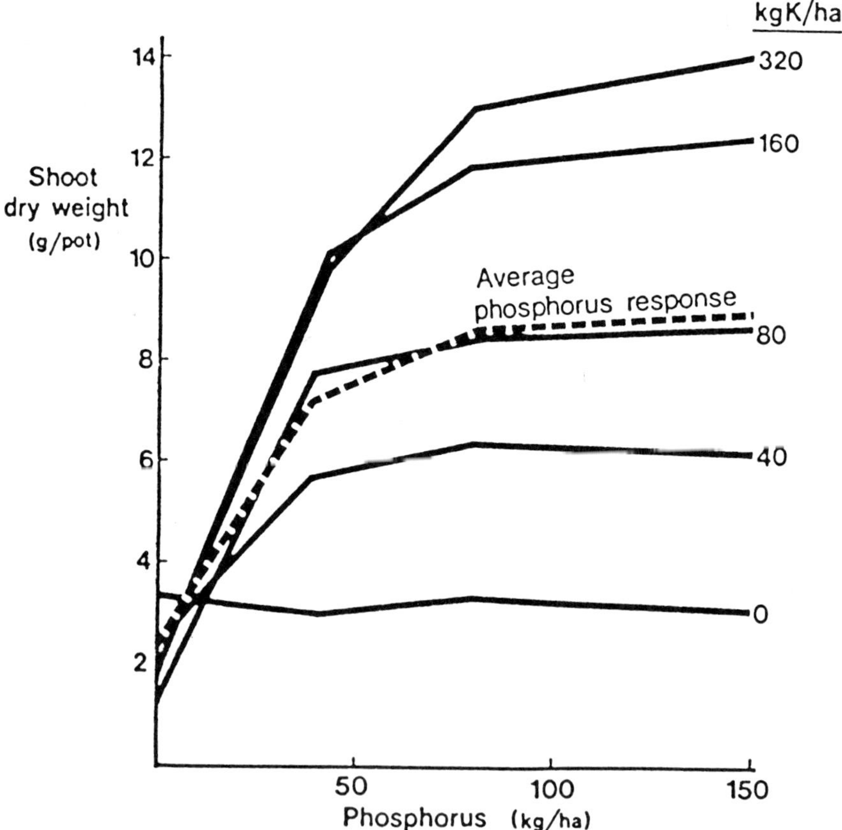

FIGURE 1 Response in shoot production of white clover growing in a deep peat hill soil to additions of phosphorus and potassium (from Rangeley, 1980).

fertiliser recommended for field use to establish white clover are 60 kg P and 80–100 kg K ha^{-1} although, as Figure 1 shows, biological responses occur to levels of fertiliser greater than this. In all experiments of this type mineral nitrogen fertiliser (Rangeley, 1980) and a cocktail of trace elements (Newbould *et al.*, 1982) were applied when responses to individual trace elements were examined at only one site was there a response to molybdenum (Floate *et al.*, 1981).

Once suitable nutrient regimes for the growth of white clover in hill soils had been determined attention was directed at the need for microbiological amendments starting with *Rhizobium* and continuing with VA mycorrhizal fungi.

Rhizobium

Scottish hill soils have few rhizobia (Holding & King, 1962) and often many of those that are present are ineffective at fixing nitrogen. Attempts to introduce effective rhizobia with the white clover seed or by spray on to the young seedling plants have had variable success, particularly on mineral hill soils (Newbould *et al.*, 1982; Mytton & Hughes, 1984). Various reasons, ranging from too much aluminium, too little water and phosphorus, the use of uncompetitive strains of *Rhizobium*, and lack of compatibility between strain and plant genotype and with soil and microclimate, have been produced to explain the situation. Workers in Northern Ireland (Wood *et al.*, 1984) have shown that aluminium only becomes a dominant influence in soils with pH less than 5.2 so that problems from this element should not be present in the studies described here since the hill soils were limed to pH 5.5 or greater.

Gibson tube tests of *Rhizobium* effectiveness were used to select a number of

TABLE 1

Response in nodule occupancy by introduced strains and in shoot production of white clover in year of sowing to inoculation of seed with *Rhizobium* (from Newbould *et al.*, 1982)

Hill soil type	Shoot dry matter (kg ha^{-1})		Nodules with introduced R. (%)	
	Inoculation			
	0	+	0	+
Brown earth	77	94	0	8
Dry peaty podsol	126	129	2	12
Wet peaty podsol	65	93	12	20
Deep peat	61	120	15	45
LSD (P<0.05)	44		7	

strains for inoculation on to white clover seeds. The peat inoculum technique with a mixed inoculum of three strains genetically marked for subsequent identification, was used in trials at 16 hill sites in the UK (Newbould *et al.*, 1982). The results for the sowing year are summarised in Table 1. Only on the deep peat soils was there a significant increase in shoot DM and this linked with the presence of the greatest number of nodules containing the largest proportion of introduced strains. The number of marked strains forming nodules was particularly low in the brown earth soils, which of all hill soils are the most suitable for hill land improvement. It was thought that the introduced strains may not have competed well with the indigenous rhizobia for nodule sites on the root.

Competition amongst strains of Rhizobium

To investigate this subject, a previously unsampled area of brown earth soil (pH 5.6) in the Cleish hills of Fife was sampled for rhizobia using the methods of Thornton (1930) and Vincent (1970). Individual strains were identified using a finger-printing technique (Josey *et al.*, 1979). Twelve individual strain families were found and their effectiveness at forming nodules, fixing nitrogen and supporting the growth of white clover seedings growing in bacteriologically controlled conditions without added mineral N are shown in Table 2. Isolates from the Edinburgh School of Agriculture and the Rothamsted collections were used for comparison. Of the twelve Cleish strains only two

TABLE 2

Herbage production ($mg\,plant^{-1}$) by white clover growing in Gibson tubes and inoculated with a range of *Rhizobium* strains from the Cleish hills, Scotland, and from the Edinburgh and Rothamsted Collections (from Jebb, 1984)

Origin	Strain Number or name	Herbage production Mean for each group		Designation of strains
Cleish	5	5.1	0.71*	Ineffective
	5	16.0	1.20	Intermediate
	2	36.0	1.56	Effective
Rothamsted	RCR10	4.0	0.60	Ineffective
	RCR221	54.0	1.73	Effective
Edinburgh	P3	41.0	1.61	Effective
	FA6	68.0	1.83	Effective
	LSD*	(P<0.05)	0.3	

*Log transformed values used for statistical analysis are underlined

could be described as effective although they produced less clover shoot dry matter than the standard Rothamsted strain (RCR 221 or TA1) and the best Edinburgh strain (FA6).

Strains of *Rhizobium* (M) were selected which were resistant to several antibiotics and were tested and found to behave similarly to the parents. These were used in competition experiments all of which were conducted under sterile conditions using the white clover cultivar NZ Grasslands Huia as host.

In paired strain experiments with 10^7 cells of each strain per ml of solution the ratio of strains occupying the nodules was directly related to the ratio of the two strains in the inoculum. Three competition indices were used and all were unaltered by changes in nutrient solution, pH or temperature. Results obtained using an equilibrium ratio method are summarised in Table 3. CL25-M, an ineffective strain, was more competitive than CL03-M and HP3-M to the extent that about three times as many cells of the latter strains would have been needed per CL25-M to result in equal numbers of nodules containing each strain on white clover roots.

Experiments were then conducted to see if priming a clover root with one strain would influence subsequent competition. Between 10^2–10^3 cells of the favoured strain were placed on the root of a seedling white clover plant. After 72 hours a 1:1 mixture of two strains with 10^7 cells ml^{-1} of each was applied to the solution bathing the roots.

The results for a comparison of uncompetitive ineffective strain CL03-M and effective strain HP3-M are shown in Table 4. Favouring the ineffective strain resulted in a dramatically smaller plant with almost 55% of the nodules formed by this strain. By contrast, a comparison of the competitive ineffective strain CL25-M with HP3-M (Table 5) showed only a modest and non-significant depression in shoot weight when the former strain was favoured. This occurred despite the presence of the ineffective strain in 42% of the nodules on the plants grown in a mixture. The reasons for the contrasted effects between

TABLE 3

The outcome of competition in bacteriologically
controlled conditions (from Jebb, 1984)

	Equilibrium ratio	Significance
CL03-M v CL25-M	0.33	P<0.05
CL03-M v CL56-M	1.0	NS
CL03-M v HP3-M	1.0	NS
CL25-M v HP3-M	3.38	P<0.05
CL56-M v HP3-M	1.0	NS

TABLE 4

The outcome of competition for nodulation of white clover when
one strain of a pair of equal competitiveness was favoured by spot
inoculation 72 hours prior to use of a mixed inoculum
(after Jebb, 1984)

	Mean shoot weight mg DM	Mean proportion of nodules that are ineffective (%)
CL03-M alone	3.6	
HP3-M	36.6	
Mixed CL03-M favoured	16.6	54.8
Mixed HP3-M favoured	41.4	31.6
LSD (P<0.05)	8.2	9.5

CL03-M is an uncompetitive ineffective strain
HP3-M is an uncompetitive effective strain

TABLE 5

The outcome of competition for nodulation of white clover when one
strain of a pair differing in competitiveness was favoured by spot
inoculation 72 hours prior to use of a mixed inoculum
(after Jebb, 1984)

	Mean shoot weight mg DM	Mean proportion of nodules that are ineffective (%)
CL25-M alone	4.1	
HP3-M	36.6	
Mixed CL25-M favoured	29.7	41.7
Mixed HP3-M favoured	22.2	29.3
LSD (P<0.05)	8.1	8.4

CL25-M is a competitive ineffective strain
HP3-M is an uncompetitive effective strain

CL03-M and CL25-M in this study are not clear and are presently being
investigated further.

The competitive nature of strain CL25-M was supported by the results of a
field experiment with white clover on a dry peaty podsol. From a mixture of
four strains inoculated on the seeds using the peat inoculum technique the
nodule occupancy (%) was as follows: 7.4 CL03-M, 16.4 CL25-M, 4.7
CL56-M (an uncompetitive effective strain) and 2.8 HP3-M. The number of
cells in the rhizosphere and bulk soil were also markedly different with HP3-M
having the least and CL25-M the most on average (Jebb, 1984).

VA mycorrhizal fungi

Many claims have been made that inoculation with VAM fungi would benefit the growth of white clover particularly in hill soils low in available phosphate (Powell & Daniel, 1978; Hayman & Mosse, 1979). Laboratory experiments at HFRO with a deep peat soil supported this (Rangeley *et al.*, 1981), but despite significant increases in the percentage of root length infected with VAM following inoculation, no significant response in shoot production was found in the field (Table 6). However, with one brown earth soil in the second year after inoculation with VAM a response in shoot production by white clover was found (Rangeley *et al.*, 1981). Field trials with white clover in New Zealand have demonstrated also the lack of consistent and predictable response to inoculation with VAM (Hall, 1984). A subsequent field experiment on a brown earth soil at Cleish in Fife, Scotland, comparing the

TABLE 6

The effect of inoculation with VAM fungi and level
of added P on production of white clover (Huia)
growing on a deep peat soil in the field and on
mycorrhizal infection (%) (from Rangeley *et al*, 1981)

	$P(kg\,ha^{-1})$				
	0		20		LSD
		Inoculation			
	–	+	–	+	(P<0.05)
Shoot DM* $(kg\,ha^{-1})$	<1[a]	1[a]	129[b]	166[b]	
Mycorrhizal infection (%)	16	53	37	52	13

*Values with different subscripts are significantly different (P<0.05)

TABLE 7

Effect of inoculation with *Glomus mosseae* L1 on root infection and shoot
growth in leeks and white clover growing in a brown earth soil in the field
(From Newbould & Rangeley, 1984)

| | Shoot production | | LSD | Root infection (%) | | LSD |
Inoculation	–	+	(P<0.05)	–	+	(P<0.05)
Plant species						
White clover $(kg\,ha^{-1})$	39	21	21.5	70	53	30.3
Leek $(g\,plant^{-1})$	14.9	22.1	6.2	46	78	19.8

effects of VAM inoculation on the growth of white clover and leeks indicated opposite responses in the two plants to the same endophyte (*Glomus mosseae* L1), production of white clover being depressed whilst that of leeks was increased (Table 7). This result suggests that best fit genotype x endophyte 'strain' relationship may occur with VAM fungi as with rhizobia (Mytton & Hughes, 1984).

INTERACTIONS BETWEEN RHIZOBIA AND VAM FUNGI

The difficulty of obtaining nodules on white clover growing in brown earth soils has already been described and the possible part played by the strain of rhizobia and the presence or absence of VAM fungi has been investigated. In particular, the effect of soil water status and/or the presence of phosphate fertiliser was explored in pot experiments with two hill soils, a dry peaty podsol and a brown earth. This experiment is described in detail elsewhere (Newbould & Rangeley, 1984) but the main results are summarised in Table 8. The presence of phosphate and water increased shoot production in both soils, depressed root infection with VAM and tended to increase nodulation,

TABLE 8

The effect of soil water status (40 or 80% water holding capacity) and amount of phosphorus (0 and 20 kg P ha^{-1}) mixed into two hill soil types on white clover herbage production (mg dry matter plant^{-1}), nodulation (No. plant^{-1}) and root infection with VAM (%) (From Newbould & Rangeley, 1984)

	Brown earth				Dry peaty podsol			
Phosphorus	0		20		0		20	
WHC (%)	40	80	40	80	40	80	40	80
Shoot DM (mg^{-1})	42	97	89	174	62	212	151	334
LSD*		38				46		
Nodules (No. plant^{-1})	0.7	1.6	1.2	1.9	25	46	58	79
LSD		0.8				26		
VAM (% root infected)	52	64	43	44	82	84	66	50
LSD		22				10		

*P<0.05

TABLE 9

The response of white clover growing in a deep peat soil
to enhanced infection of the roots with VA mycorrhizal
fungi (from Newbould & Rangeley, 1984)

| | Inoculant | | LSD |
	−	+	(P<0.05)
Root infection (%)	13	54	5.0
Plant DM (Mg plant^{-1})			
Shoot	18	45	6.1
Root	8	20	3.2
Total	26	65	8.7
Nodules (No. plant^{-1})	5.5	12.7	2.18
C_2H_2 reduction (n moles plant^{-1} hour^{-1})	50	475	164
N in shoots (mg plant^{-1})	1.0	1.8	0.25

though this was significant in the dry peaty podsol soil only. The marked contrast in nodulation between white clover plants grown in the brown earth and dry peaty podsol is very evident.

Despite the lack of linkage between the two micro-organisms shown in the data of Table 8 there are other indications of a double symbiosis. A pot experiment with white clover grown in a deep peat soil shows the benefit in plant growth, nodulation, nitrogen fixation and nitrogen content of shoots that can follow an increase in root infection with VAM (Table 9). In this trial it was evident that the uninoculated plants were short of phosphorus and an increase in the proportion of root infected with VAM fungi following inoculation was able to increase uptake of this element. The presence of extra phosphate apparently permitted the rhizobia to express themselves by increased nodulation and fixation of nitrogen. The presence of this double symbiosis and the knowledge that host genotype must be matched with *Rhizobium* strain, and possibly with VAM endophyte, indicates that it may be necessary to ensure that microbes for use in field inoculants must be compatible with each other, with the host and with the environment. Failure to achieve the perfect blend in the past may explain the lack of consistent and predictable responses to microbial inoculation in hill soils.

DISCUSSION

From the evidence briefly reviewed here, it is apparent that the potential of white clover to grow and fix N in Scottish hill soils can be limited by lack of appropriate and compatible micro-organisms. More basic knowledge of both rhizobia and VA mycorrhizal fungi is required before consistently practicable and effective inoculants can be produced. For rhizobia, it is suggested that elite strains which are both competitive and effective at fixing N are required and procedures are available which might result in their production. However, even given these elite strains, further understanding of the ecology of rhizobia in soil, of recognition, infection, nodule growth and N fixation processes is needed before practically applicable inoculation methods for their use can be designed.

Similarly, much more information is needed on the autecology of VA mycorrhizal fungi and on the processes of infection and the physiology of their function before practically applicable inoculation procedures can be devised for use in the field. Studies on this micro-organism would be helped if methods to identify endophytes and to culture them axenically could be formulated.

Suggestions that the two micro-organisms can interact to their mutual benefit, and to that of their host plant are interesting and worthy of further study. Likewise the need remains to match inoculant *Rhizobium* strain and/or endophyte to host plant genotype and to soil and environmental conditions, and much work is needed to ensure consistent and reliable responses in

TABLE 10

Microbes and the growth of white clover in hill soils.
Conclusions and the way forward . . .

Conclusions

Rhizobium strains differ in the ability to form nodules and to fix nitrogen, and can show host genotype preferences. Response to inoculation in the field is inconsistent, and especially so in brown earth soils.

VA mycorrhizal fungi can enhance nodulation, nitrogen fixation, phosphorus uptake, and shoot growth of white clover in the laboratory, but responses to inoculation in the field are unpredictable. There is some evidence of endophyte × host genotype interactions.

The Way Forward

Rhizobium. Select *Rhizobium* strains for competitiveness and persistence in soil and attempt to transfer N-fixing ability by microbial genetic techniques. Increase understanding of ecology, recognition, infection, nodule growth, nitrogen fixation, and the metabolism processes of fixed nitrogen in plants.

VA mycorrhizal fungi. Increase understanding of process of infection and of functional physiology of both indigenous and introduced endophytes. Devise methods to grow axenically and to identify endophytes.

Develop practically applicable inoculation procedures.

---

practice. The conclusions from work with microbes and white clover at HFRO are summarised in Table 10 together with possible areas for further study. It is disappointing in many ways that, although marked responses to inoculation with both *Rhizobium* and VAM were found in laboratory experiments, inoculation of white clover in the field has yielded such inconsistent results. It appears that inoculants and inoculation procedures have been used before the basic biology and ecology of the organisms and their relationships with host plants and the soil and climatic conditions of bulk soils and pastures are fully understood. More basic research is required before practically applicable and farmer acceptable procedures can be formulated.

The engineering or selection of appropriate microbial symbionts and the development of methods of inoculation practicable for routine field use should enhance the growth of white clover in hill pastures. In turn this should help to increase production of high quality herbage which, if used strategically in altered traditional management systems, should increase the output of ruminant animals from the hills and uplands. Use of effective microbial symbionts will lessen the need for expensive nitrogen fertiliser and reduce the level of phosphorus fertiliser needed. However, it should be noted that this impact of microorganisms on legume growth and function, and on agricultural production from the hills and uplands, cannot be achieved without the concomitant application of lime, phosphorus and potassium. Thus, it is necessary to retain a balanced approach to the improvement of hill pastures and to emphasise that a combination of both microbial and chemical additives is required.

SUMMARY

Hill soils are acid, low in available plant nutrients, especially N, P and K, high in organic matter, and generally low in microbiological activity.

Improved quality of herbage is needed to increase the economic viability of sheep production from hill pastures. This can be obtained by ameliorating the soils with lime and fertilisers and sowing grass and white clover seeds. White clover is desirable both because of the high quality of its herbage for grazing livestock and its ability to fix atmospheric nitrogen. Despite the presence of low numbers of only moderately effective *Rhizobium* sp. in many hill soils responses by white clover to inoculation of the seeds with selected effective strains have been inconsistent. Lack of competitiveness in the introduced strain is shown to be a possible reason for the poor response to inoculation, especially in mineral soils. Similarly, increases in the proportion of clover roots infected with VA mycorrhizal fungi following field inoculation are not reflected in enhanced uptake of P and growth of the plants as observed under laboratory conditions. Some evidence for the benefits of double symbiosis between *Rhizobium* sp. and VA mycorrhizal fungi and white clover plants has been obtained. Recent work at HFRO on the responses of white clover growing in hill soils to inoculation with both *Rhizobium* sp. and VA mycorrhizal fungi, and some of the reasons for lack of predictability in field response, are described and discussed.

References

Floate, M.J.S., Rangeley, A. & Bolton, G.R. (1981). An investigation of improved pasture production on deep peat with special reference to potassium responses and interactions with lime and phosphorus. *Grass & Forage Science,* **36,** 81–90.

Hayman, D. & Mosse, B. (1979). Improved growth of white clover in hill grasslands by mycorrhizal inoculation. *Annals of Applied Biology,* **93,** 141–148.

Hall, I.R. (1984). Field trials assessing the effect of inoculating agricultural soils with endomycorrhizal fungi. *Journal of Agricultural Science, Cambridge,* **102,** 725–731.

Holding, A.J. & King, J. (1963). The effectiveness of indigenous populations of *Rhizobium trifolii* in relation to soil factors. *Plant & Soil,* **18,** 191–198.

Jebb, J.R. (1984). Competition between strains of *Rhizobium trifolii* for nodulation of white clover (*Trifolium repens*). Ph.D. Thesis, University of Edinburgh.

Josey, D.P., Benyon, J.L., Johnston, A.W.B. & Beringer, J.E. (1979). Strain identification in *Rhizobium* using intrinsic antibiotic resistance. *Journal of Applied Bacteriology,* **46,** 343–350.

Maxwell, T.J., Eadie, J. & Sibbald, A.R. (1979). The economic implication in the application of new techniques to hill sheep farming in Scotland. In *Hill Lands: Proceedings of an International Symposium* (J. Luchok, J.D. Cawthorn & M.J. Breslin, eds.), pp. 306–311. West Virginia University Books; Morgantown, VA.

Mytton, L.R. & Hughes, D.M. (1984). Inoculation of white clover with different strains of *Rhizobium trifolii* on a mineral hill soil. *Journal of Agricultural Science, Cambridge,* **102,** 455–459.

Newbould, P., Holding, A.J., Davies, G.J., Rangeley, A., Copeman, G.J.F., Davies, A., Frame, J., Haystead, A., Herriot, J.B.D., Holmes, J.C., Lowe, J.F., Parker, J.W.G., Waterson, H.A., Wildig, J., Wray, J.P. & Younie, D. (1982). The effect of *Rhizobium* inoculation on white clover in improved hill soils in the United Kingdom. *Journal of Agricultural Science, Cambridge,* **99,** 591–610.

Newbould, P. & Rangeley, A. (1984). The effect of lime, phosphorus and mycorrhizal fungi on growth, nodulation and nitrogen fixation by white clover (*Trifolium repens*) grown in U.K. hill soils. *Plant & Soil,* **76,** 105–114.

Powell, C.Ll. & Daniel, J. (1978). Growth of white clover in undisturbed soils after inoculation with efficient mycorrhizal fungi. *New Zealand Journal of Agricultural Research,* **21,** 675–681.

Rangeley, A. (1980). The nutrient requirements of white clover on hill soils. Ph.D. thesis, University of Edinburgh.

Rangeley, A., Daft, M.J., & Newbould, P. (1981). The inoculation of white clover with mycorrhizal fungi in unsterile hill soils. *New Phytologist,* **92,** 89–107.

Thornton, H.G. (1930). The early development of the root nodule of lucerne (*Medicago sativa* L.). *Annals of Botany,* **44,** 385–392.

Vincent, J.M. (1970). *A Manual for the Practical Study of Root Nodule Bacteria,* IPB Handbook No. 15. Blackwell Scientific Publications; Oxford.

Wood, M., Cooper, J.E. & Holding, A.J. (1984). Soil acidity factors and nodulation of *Trifolium repens. Plant & Soil,* **78,** 367–379.

Ecosystem Manipulation for Increasing Biological N$_2$ Fixation by Blue-Green Algae (Cyanobacteria) in Lowland Rice Fields

I.F. Grant[1], P.A. Roger[2] and I. Watanabe

The International Rice Research Institute, Los Banos, Laguna, Philippines

INTRODUCTION

Lowland rice is grown under flooded conditions. Submergence is of considerable importance for the maintenance of soil N fertility while regularly cropping rice without nitrogenous fertiliser. In these conditions the natural supply of N derived from irrigation water and rain is generally small (Yamaguchi, 1979) in comparison with that supplied by biological N$_2$-fixation (BNF). The latter, which is estimated from N$_2$ balance studies as 15–50 kg N crop^{-1} (Koyama & App, 1979) has therefore been designated responsible for the continued maintenance of the crop yields. De (1936), attributed much of this natural fertility to N$_2$-fixing blue-green algae (BGA) and Okuda (1948) and Konishi & Seino (1961) found a correlation between the growth of algae and soil N gain. Increasing BNF is clearly desirable for resource neutral agricultural systems, particularly as the cost of commercial fertiliser is unpredictable.

Methods to increase biological N$_2$-fixation by algae in rice fields have included the use of P$_2$O$_5$, NaMoO$_4$ and lime as soil amendments, and these were measurably successful (De & Mandal, 1958, Amma *et al.*, 1966). Encouraging results were also obtained using the algal inoculation technique (Watanabe *et al.*, 1951; Singh, 1961; Subrahmanyan *et al.*, 1965; Aboul-Fadl *et al.*, 1967; Venkataraman, 1972) that relies upon the success of introduced strains of blue-green algae to fix atmospheric N$_2$. The technique is often referred to as a small-scale village biotechnology. However, many failures are

[1]Boyce Thompson Institute for Plant Research, Cornell University, Ithaca, NY 15853, U.S.A.
[2]Maitre de Recherches ORSTOM (France), visiting scientist at IRRI.

reported from different geographical locations; Japan (Watanabe, 1973); Taiwan (Huang, 1978); Philippines (Alimagno & Yoshida, 1975) and India (All India Coordinated Project on Algae, 1979). Moreover, nurturing inoculum growth was often difficult (Konishi & Seino, 1961). Competition from indigenous algae (Yamaguchi, 1979) and grazing of inoculum by invertebrates (Hirano *et al.*, 1955; Watanabe *et al.*, 1955) were detrimental to inoculum establishment. The role of some invertebrate grazers in restricting the growth and nitrogen fixing activities of BGA has been evaluated by Wilson *et al.* (1980), and Osa-Afiana & Alexander (1981); and consumption of BGA, quantified by Grant & Alexander (1981) and Grant *et al.* (1983a). These results collectively support the notion that primary consumers that proliferate at the expense of BGA are a major limiting factor to algal N_2 fixation in lowland rice fields. This paper will introduce the rice field ecosystem with emphasis on the grazing components and then will focus on the recent attempts made to increase the contribution of N_2 fixed by both indigenous and inoculated BGA through the manipulation of the rice field ecosystem.

LOWLAND RICE ECOSYSTEM

In general, the flooded rice field is an uncompromising habitat. With few exceptions, lowland rice fields are temporary aquatic environments that are subject to physical and chemical extremes of insolation, temperature, pH, O_2 concentration and nutrient status. Furthermore, ploughing, transplanting and weeding are cultural practices disruptive to the establishment of community stability. A rather specialised rice field fauna and community structure might therefore be expected to result. Indeed, rice cultivation freezes ecosystem development as a secondary succession, and thereby prevents the ecosystem from reaching any natural conclusion or climax. Presumably, if cultivation were suspended, rice fields would revert back to a marshland community.

Although continuous, lowland rice field habitats may be broadly distinguished into 4 major zones:

Floodwater

This is the photosynthetic (photic) zone when light penetration is not impaired by turbidity caused by rice cultural practices (land preparation, transplanting and weeding) heavy rain or the activity of benthic invertebrates which stir up clay and silt particles. Floodwater supports the photosynthetic and chemosynthetic producers (algae and bacteria), invertebrate and vertebrate primary consumers (grazing zooplankton) and the nektonic secondary and tertiary consumers (carnivorous insects and fish). The distribution of aquatic flora and fauna is probably influenced by the chemistry of the soils, so that biocoenoces characteristic of neutral, acid sulphate, alkaline, saline and peat soils should be evident. As yet, qualitative and quantitative observations are too scarce to permit comparisons.

Data on oxygen, temperature and even major limiting nutrients, e.g., N, P, Ca, are also of isolated occurrence, and data collection over an ecologically useful period of time, such as a cropping season, has not been maintained. Some diurnal curves are available for rainfed (Heckman, 1979) and irrigated fields (Saito & Watanabe, 1978) which show, at least in the presence of submerged weeds, dissolved oxygen to range between daytime supersaturation and anoxic conditions at night. Partial pressures of CO_2 are inversely proportional to pO_2 and consequently diurnal pH change from a normally neutral floodwater to pH 9.5 is not uncommon at times of algal blooms or luxuriant weed growth. Floodwater temperature is frequently subjected to a 10°C change diurnally and maxima during mid-afternoon, measured at the soil/floodwater interface, often reach 36–40°C.

Except for nutrient flux calculations following fertiliser applications (De Datta *et al.*, 1983), measurements of mineral nutrients in floodwater have not been attempted routinely. Release of nutrients into the floodwater after land preparation, particularly following dry fallowing, is certainly rapid for mineralised N (Shiga & Ventura, 1976) and probably accounts for the initial bloom of algae frequently observed about 2 weeks after puddling (Kurasawa, 1956; Saito & Watanabe, 1978).

Surface soil

By virtue of the maintenance of shallow floodwater (2–10 cm), the surface soil has a redox potential higher than 300 mV. The actual depth of this oxidised layer is dependent upon the dissolved oxygen concentration of the floodwater, the reducing capacity of the soil, and the activities of the benthos and infauna. It is usually between 2 and 20 mm (Watanabe & Roger, 1984) and distinguishable visually by colour and chemically by the predominance of ferric over ferrous ions. After land preparation, the surface soil is a site of algal growth, which supports a large grazing population. Later in the rice growing season, decomposable organic matter accumulates in this zone from which benthic filter and deposit feeders (e.g. rotifers and molluscs) acquire their energy. The activities of benthic invertebrates affect nutrient recycling either directly by excretion or indirectly by release of native minerals through soil perturbation. The surface-oxidised zone is microbiologically active. Aerobic and microaerobic conditions favour microbial processes which may affect the efficiency of N cycling (nitrification and N$_2$ fixation).

Reduced zone

The anaerobic reduced zone soil underlying the oxidised layer is frequently called the reduced layer. The Eh range is 300mV to -300mV. Reduction processes predominate and NH_4-N, sulphide and methane and organic acids are liberated. Decomposing organic matter in the reduced zone sustains populations of tubificid worms and chironomid larvae whose burrowing activities aid ammonia diffusion (Fenchel & Blackburn, 1979; Kikuchi & Kurihara, 1982; Gardner *et al.*, 1983; Grant & Seegers (in press)) and PO_4 release into the floodwater (Gardner *et al.*, 1981). Animals inhabiting this zone frequently contain haemoglobin or possess air sacs as adaptions to low oxygen concentrations.

Rice plant

The rice plant creates conditions which modify the soil/floodwater environment. Shading of the floodwater and surface soil increases as the rice canopy enlarges. The ensuing changes in light intensity affects the growth of photodependent organisms. In Senegalese rice fields, BGA growth was favoured by lowered light intensities (Reynaud & Roger, 1978) but not all BGA are distributed in relation to high or low light intensities. Rice also indirectly affects the floodwater/soil communities by lowering the temperature and CO_2 concentrations under the canopy. Reduction of solar radiation coupled with low CO_2 levels on calm days will affect growth rates, succession and perhaps distribution of autotrophic organisms. Rice plants act as substrate for epiphytic growths (Roger *et al.*, 1981) and provide mechanical support for many animal species. Pulmonate molluscs may escape high floodwater temperatures by resting at the air/floodwater interphase.

Rice roots remove nutrients from the floodwater and soil and exude organic substrates utilisable by bacteria as carbon and energy sources. Oxygen translocated from the aerial parts of the rice plant to the roots oxidises the rhizosphere and may create conditions favourable for the growth of N_2-fixing organisms (Watanabe & Roger, 1984).

GRAZING COMPONENT

Animals that obtain all or part of their energy requirements by filtering, browsing or shredding primary production are called grazers. As there is a close relation between algae and grazers early in the rice cultivation cycle (autotrophic succession), the possibility exists to relieve grazing pressure and encourage growth of N$_2$-fixing algae. We shall therefore concentrate on this part of the ecosystem.

Copepods, cladocerans and rotifers are typical planktonic grazers which filter phytoplankton and bacteria from the floodwater. At the floodwater/soil interphase, ostracods, chironomid larvae and molluscs browse the algal growths which luxuriate early in the rice cropping cycle. Quantitative studies of zooplankton and benthic invertebrates have almost exclusively been made in fertilised fields and the units of density are not usually comparable. However, Kurasawa (1956) recorded, for example, 198 *Daphnia*, 15 *Bosmina*, 42 *Cyclops*, 42 *Branchionus* and 56 *Keratella* (per litre) about 6 weeks after planting. Crustacean zooplankton densities ranged between 200–800 l^{-1} in another Japanese study (Kikuchi *et al.*, 1975). Many of these small filter feeders are not able to feed upon the large filamentous BGA, which are rejected by their feeding apparatus. The larger benthic invertebrates can readily ingest the N$_2$-fixing algae and large populations of ostracods (10–20,000 m^{-2}) chironomid larvae (8000 m^{-2}) and molluscs (up to 1000 m^{-2}) have been observed in Philippine rice fields (unpublished).

By combining the data of Pantastico & Suayan (1973), Watanabe *et al.* (1978), and Saito & Watanabe (1978) with our own observations, a generalised succession of the predominant dry season flora and fauna from unfertilised rice fields of the Philippines is presented in Figure 1. Colonisation of the prepared and puddled rice field is rapid unless the land was not previously flooded. Propagating bodies (such as spores, akinetes etc.) quickly produce unicellular eukaryotic algae which are soon succeeded, albeit briefly, by fast-growing heterocystous BGA about 2–3 weeks after transplanting. The chlorophyll a concentration of the floodwater and floodwater/soil interphase reaches a peak as a result of this BGA bloom, but rapidly subsides to a basal level (0–3 mg Chl.a m^{-2}) thereafter. Fields of higher natural fertility will possess greater concentrations of chlorophyll a or chlorophyll-like substances (Wada *et al.*, 1982). About 4 weeks after transplanting the population of the floating and slower growing mucilagenous colonial BGA develop, and these are observed to reach a maximum biomass just before the rice harvest. A cause of the decline in the initial bloom of BGA is thought to be grazing by ostracods and molluscs. Flooding induces rapid ostracod development from eggs resistant to desiccation. The young increase their population density at the expense of bacteria and later, when their size allows it, the BGA. The collapse of the ostracod population quickly follows that of the BGA population, i.e., about 4 weeks after transplanting.

The recruitment rates of molluscs are dependent upon the previous condition of the rice field. When flood fallowed or wet, molluscs may swiftly attain population densities detrimental to BGA growth, but these populations quickly subside with the disappearance of algae, which also signals the change from an autotrophic to an indicated heterotrophic succession.

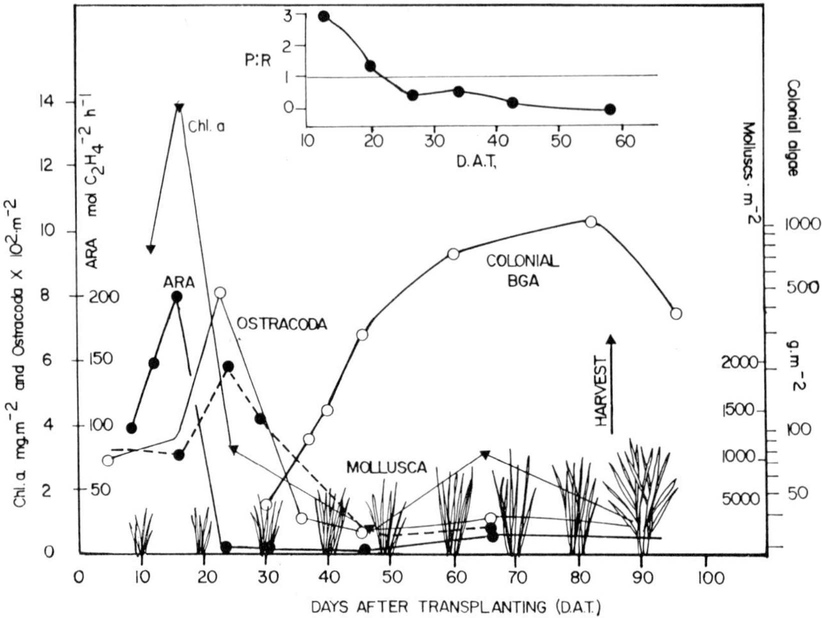

FIGURE 1 Generalised fluctuations of algae, ARA and grazers in floodwater of unfertilised lowland rice fields. Production: respiration (P:R) calculated from Ichimura (1954).

A heterotrophic succession is only indicated from data using O_2 measurements of productivity. Underestimates of O_2 concentrations due to non-biological O_2 consumption (Saito & Watanabe, 1978) and cultural practices such as removal of filamentous algae or submerged weed may cloud the real picture. Bacterial numbers are greatest following algal growth.

Primary production typically exceeds community respiration (P:R > 1) over the first month of rice cultivation (inset, Fig. 1) which leads to the build up of organic matter and a decrease of the ratio to below 0.1. The instability of the rice field ecosystems prevents maturity, which would lead to reduced grazing through increased food chain complexity.

Coincident with the initial BGA bloom, which is not usually resistant to grazing, is a peak of nitrogen fixation activity. Measured by the acetylene reduction assay (ARA), which usually underestimates the actual fixation rate, this was equivalent to about 2–3 kg N ha⁻¹ over its duration (18 days). Arriving at a meaningful figure of N_2-fixation by the floating colonies of BGA later in the season is difficult but, Watanabe *et al.* (1978) present data from which it might be extrapolated. However, even if fixation of the late season BGA is considerable, the efficiency of N transfer to the succeeding rice crop must be low but there is no data in support of this. For this reason, regulation of a grazing population to increase initial growth and N fixation of BGA should be attempted. Fixed nitrogen released from algal cells would then become available at times of the rice N requirements, i.e., tillering and panicle initiation of the planted crop. Moreover, rice inoculation techniques are usually implemented immediately after land preparation, so techniques to improve survival of inocula such as a reduction of grazer pressure, need implementation at the same time.

EFFECTS OF CONTROLLING GRAZING

In field experiments conducted at the International Rice Research Institute (Grant *et al.*, 1983b), grazing by Ostracoda was arrested using Perthane or crushed neem seeds (*Azadirachta indica*). The active principals in neem seeds reduce ostracod feeding and growth (Grant *et al.*, 1984). Molluscs, the other dominant grazer of BGA at this site, were controlled with Bayluscide. Four applications of these control agents were made at two week intervals from transplanting. Populations of ostracods (Fig. 2a) and molluscs were effectively controlled until ripening stage by neem and Perthane + Bayluscide (P + B) applications. Population densities of ostracods rapidly increased in control plots.

Heterocystous BGA were dominant in all plots from transplanting to 59 days after transplanting (DAT) when chlorophytes succeeded. Throughout the cultivation cycle, chlorophyll a concentrations were significantly higher in those treatments which reduced ostracod populations (Fig. 2b). *Nostoc* sp. and *Anabaenopsis* sp. were characteristic of the plots with lowered grazer pressure and blooms of these algae lasted about 30 days. From 50 DAT onwards the dominant BGA were mucilagenous colonial strains of *Gloeotrichia, Nostoc* and *Aphanothece*. Early in the cultivation cycle, the blooms of algae in neem and P + B treatments fixed significantly higher amounts of N (as measured by ARA, Fig. 2c), than the control (10 times less). As a result of higher N$_2$-fixation rates attributed to ostracod population control, grain N was significantly increased by 37% and 24% compared with the controls using neem and P + B respectively. It was noted in the case of neem treatments, where *Anabaenopsis* bloomed at 31, 59 and 72 DAT in replicates, that grain yield and grain N was highest in that replicate which bloomed 31 DAT. It was speculated that algal N available at this stage contributed to yield increases whereas later blooms could only contribute to successive crops.

The application of neem seeds or Perthane + Bayluscide therefore extended the length and increased the intensity of the BGA blooms and hence the N$_2$-fixation capacity. (cf with natural conditions in Fig. 1).

These results, particularly with neem, which is a cheap alternative to

OSTRACODA No./250 ml

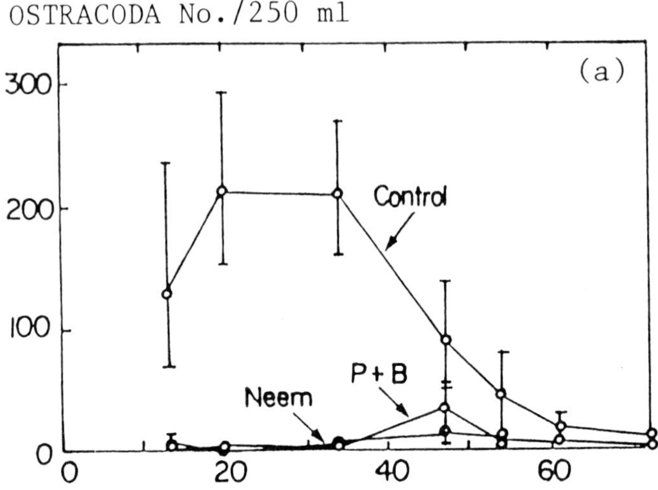

Chlorophyll a mg m^{-2}

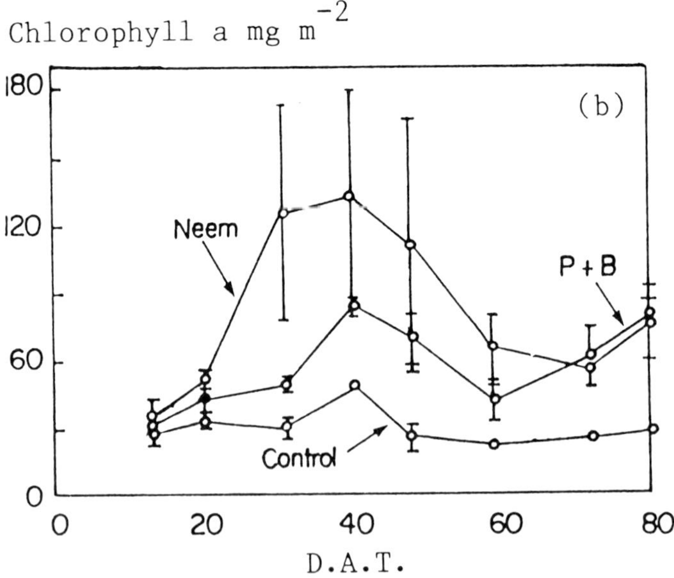

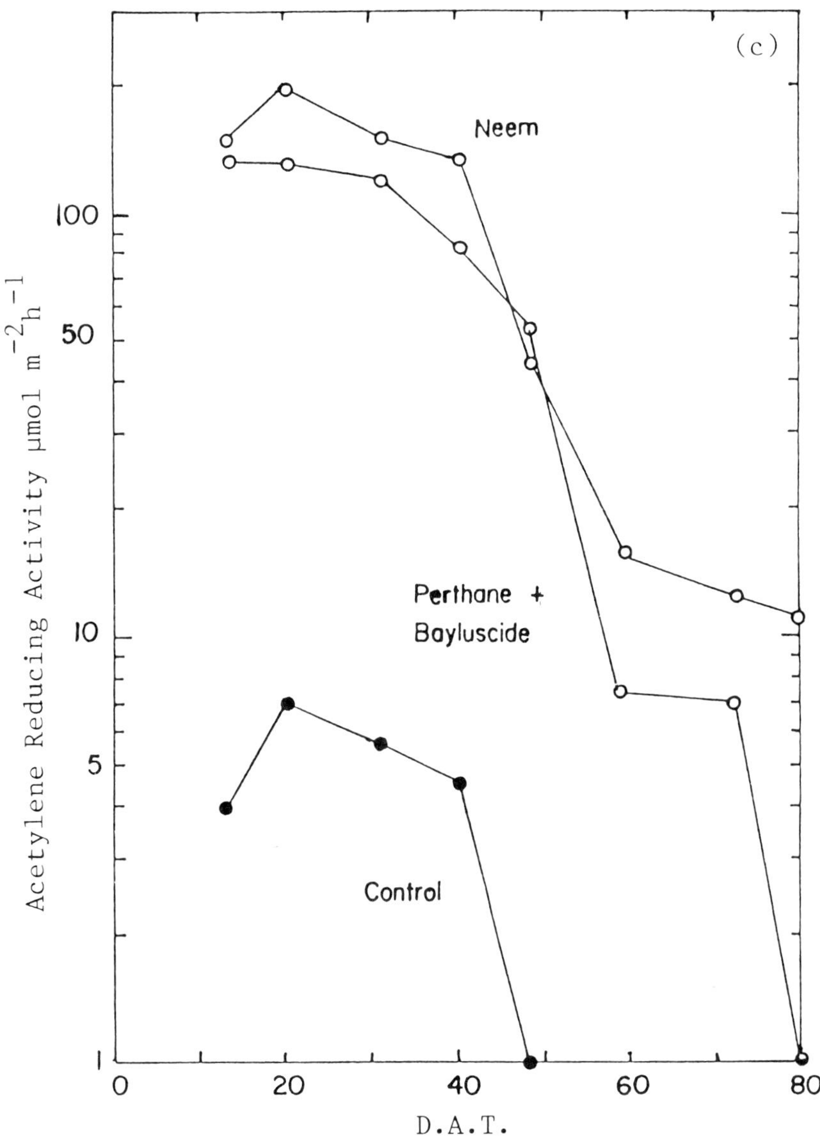

FIGURE 2 a) Mean density of ostracods ($\pm$ S.E.), b) Mean chlorophyll concentrations and c) sliding averages of total ARA. (Floodwater+soil surface ARA) in three treatments. DAT=Days after transplanting.

TABLE 1

Nitrogen balance after five crops

			mg N			
Treatment	Crop	Initial soil	Final soil	N inputs	N balance[a]	N gain as % of crop N
Neem seed	421±39	4215± 65	4444±49	40	609[a]	144
Neem cake	403±48	4257± 43	4218±82	82	280[bc]	69
Control	382±13	4369±138	4238±89	0	251[bc]	65
Neem seed, black cloth	249±21*	4219± 64	3988±46**	40	22[c]	7.4

[a]Mean±S.E. Means followed by the same letter are not significantly different at 5% level
*, **Significant at 5 and 1% level

conventional pesticides, were encouraging, but it was not certain that reduction of grazing and thus increased N_2-fixation by BGA was responsible for the increases in rice protein. A N balance study by Grant *et al.* (1984) quantified the amount of N_2 fixed biologically by BGA as a result of one application of neem. Pots containing flooded soil were inoculated with 20 ostracods and a soil slurry containing BGA. Crushed neem seed and powdered neem cake (at 100 and 200 kg ha^{-1} respectively) were administered to control ostracod growth. A neem seed treatment with a black cloth which covered the pot prevented photodependent N_2 fixation. The difference between initial and final soil N content plus rice crop N uptake is the N_2 balance, any N inputs such as N in neem, having been subtracted. After 5 crops a large and significant N balance was obtained in neem seed treated plots (Table 1). This was ascribed to increased N_2 fixation by BGA as seen from that small balance obtained in the absence of light (i.e. heterotrophically). Crop N uptake was similar between treatments except when BGA were excluded. Thus the large N surplus accrued in neem seed treated soils was either unavailable to the rice (immobilised) or was due in part, to less N_2 loss. The latter is possible as neem is a known nitrification inhibitor (Bains *et al.*, 1971). Neem cake was not effective in producing an N surplus.

GRAZER CONTROL WITH INOCULATION

A field experiment in the Philippines again employed neem to control ostracod grazing while an algal inoculum, composed of 2 fast-growing and efficient N_2-fixing strains, were broadcast onto experimental plots. In order to allow the algae to establish while the grazers were suppressed, an application of neem was made at the beginning of the experiment only a few days before

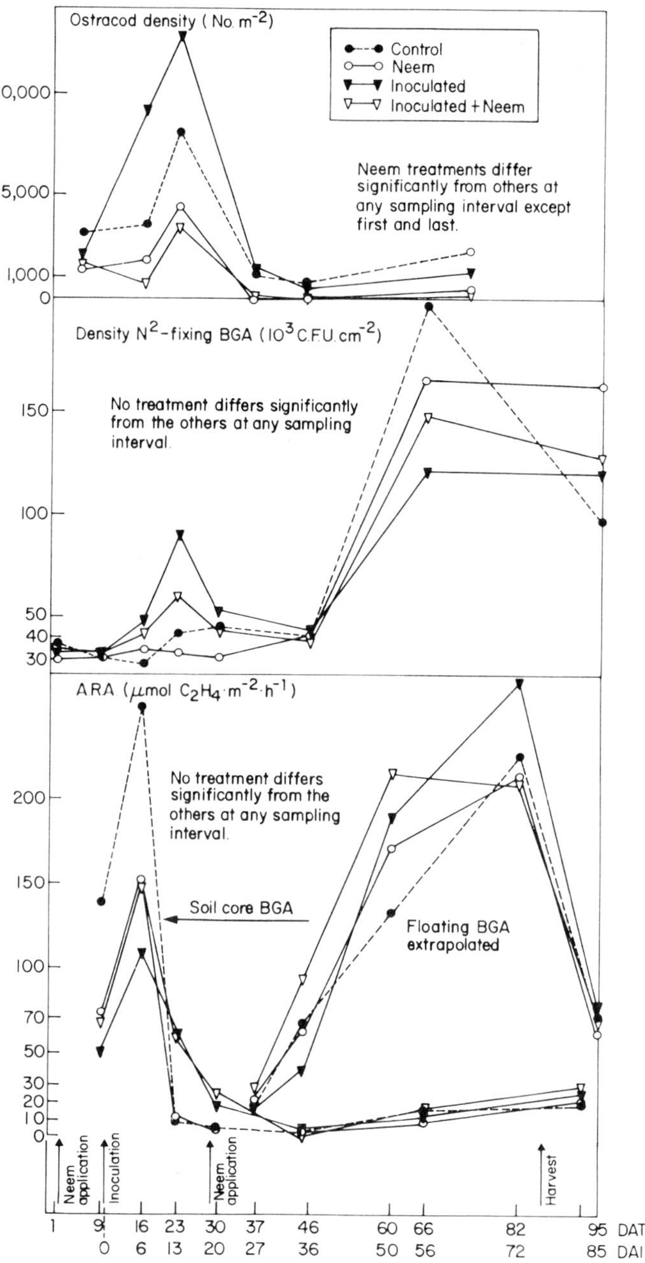

FIGURE 3   Ostracod density, density of BGA and ARA in algal inoculation trial. DAT = days after transplanting. DAI = days after inoculation.

the powdered algal inoculant was administered. Plots were treated with 1) a combined *Tolypothrix tenuis* and *Nostoc* sp. inoculum; 2.5 kg dry wt. ha^{-1}, 2) crushed neem seeds, 3) inoculum + neem seeds and 4) a control receiving no treatment. All received P as 5 kg KH$_2$PO$_4$ ha^{-1}.

Ostracods reached their maximum density in inoculated plots (Fig. 3). Both neem treatments contained significantly less dense populations of ostracods than the control and inoculated plots. All ostracod populations crashed 4 weeks after inoculation, presumably as food became exhausted.

Floodwater and soil samples were plated upon a N-free agar growth medium. Figure 3 shows the density of colony-forming units of N$_2$ fixing BGA (CFU cm^{-2}) from transplanting to post harvest drying of the field. There were no significant differences of CFU between treatments and the tendency towards more CFU in inoculated and inoculated + neem treatments just 2 weeks after inoculation was not due to inoculated strains. Inoculated strains were not recorded. Coinciding with the ostracod population crash one month after inoculation was an increase in CFU of *Nostoc* spp. evidently derived from floating mucilagenous strains which are highly resistant to grazing. This floating biomass consisted of *Gloeotrichia, Nostoc* and *Aphanothece* and no differences in their weight between treatments were evident. Twelve to 15 t ha^{-1} were recorded (fresh wt), but this was only equivalent to 2.7 kg N ha^{-1}.

The characteristic early burst of N$_2$-fixing activity corresponding to the fast growth of indigenous strains of *Anabaena* (Fig. 3) occurred in all treatments. Significant differences in ARA were not apparent. The second peak (Fig. 3) is extrapolated from laboratory assays of harvested mucilagenous BGA and, as such, is difficult to interpret. There were no differences in grain yield or grain N between treatments.

In the second crop, neem seeds of different origin were used. As a result, better control of ostracods was achieved and ARA was doubled, but again the inocula were not responsible for the increased fixation. Rather, blooms of indigenous *Anabaena* sp. were identified as responsible.

GRAZER CONTROL AND ALGAL INOCULATION IN SUSTAINABLE AGRICULTURE

Sustainable agriculture usually requires low cost inputs and practices. Application of the seeds of *Azadirachta indica* provided control of ostracod grazing and permitted increases in algal biomass, N$_2$ fixation and grain N. However, variation in the effectiveness of neem seeds according to origin and storage conditions has been reported (Schumtterer & Zebitz, 1984), and this was considered responsible for a less than satisfactory control of ostracods in the first inoculation experiment. The efficiency and economy of neem in increasing BNF may be maximised by proper timing of applications to coincide with transplanting or broadcasting of rice and phosphorus application. In the field, neem will adequately control ostracod populations for up to 18 days.

Grazing of BGA by molluscs assumes importance if land preparation is done much in advance of transplanting and P application, or if floodwater is obtained from irrigation canals. As much sustainable rice agriculture relies upon rain, simple cultural procedures such as planting and P application soon after ploughing can alleviate the problem of grazing from molluscs. Alternatively, the seeds of berries of *Phytolacca dodecandra, Croton tiglium, Jatropha curcas* (McCullough *et al.*, 1980) and crushed cashew nut shells (*Anacardium occidentalis*) are also molluscicidal (Webbe & Lambert, 1983) and offer economical substitutes for expensive commercial preparations.

Plants having pesticidal properties are frequently cheap and may be free if grown by the farmer. Many plants such as neem and those listed with molluscicidal properties are frequently available in rice growing areas. Extracts of seeds and leaves are often more efficient and effective than the addition of whole plant parts, and they may be made with simple equipment. When grazer control measures were extended over 6 weeks from transplanting the short-lived BGA population normally observed early in the rice cultivation cycle was extended over a 40 day period. Repetition of treatments to gain longer control of grazers and more intense BGA growth was costly but resulted in grain N increases unlike simple applications of neem. Control measures administered only once (inoculation trials) at transplanting showed a positive effect upon N$_2$-fixation but the increase was only a small proportion (5%) of the total N$_2$-fixing activity over the whole rice cropping cycle (e.g. 15–50 kg N ha^{-1}, Koyama & App, 1979). Control measures maintained over 6 weeks increased the proportion of N$_2$-fixed to approximately 10% of the total N fixed per crop. Under stable conditions of the greenhouse, the N balance experiment showed that N fixed as a result of a short period of grazer control was considerable (144% N gain as % of crop N), but much of the N which accrued in the soil was not used by the crop. The N balance experiment, clearly showed the potential of BGA in submerged soil but the reasons for N immobilisation need elucidation. Furthermore, any effects of neem upon prevention of N losses by nitrification cannot be contrasted against the N gains by N$_2$ fixation. It is too early to make decisions using economics as the criterion for length of grazer control. Only further field trials can establish how long treatments should be and much variation is to be expected between rice fields. Whether such manipulative practices are economically feasible may depend upon the region.

Another approach to grazer control is the development of grazer-resistant algae. It has been noted already that many mucilagenous BGA strains have extensive resistance to invertebrate grazers. However, the mucilagenous strains are slow growing and thus characterise the later stages of rice cultivation. Their biomass may be high; up to $2.4\,kg\,m^{-2}$ (wet weight) has been reported in the Philippines, but this is exceptional and moreover only represents a small amount of N (viz. $2\,g\,N$ and $0.7\,g\,P\,m^{-2}$ as $N = 3\%$ and $P = 1\%$ of dry wt. of alga: IRRI, 1983). Mucilagenous strains generally show much less ARA than the fast-growing non colonial BGA typical of the early part of rice cultivation. Strains which have doubling time of 8h are on record (Chen, 1983) but very few of the faster growing and efficient N_2 fixing strains have shown resistance to ostracod grazing, and those which have were of temperate origin (Grant, unpublished). Screening for resistance is continuing because a successful candidate would be a better alternative economically, than natural pesticide control methods. Nevertheless, the problems associated with nurturing algae to establishment are apparently great and maintenance of a bloom difficult because other grazers gradually exploit the available energy source. A better knowledge of the soil/floodwater ecosystem would increase the chances of successful inoculation, but there has been little advance in our understanding of the ecosystem since attention was drawn to this scarcity of information by Roger & Kulsaooriya (1980).

Inoculation of non-indigenous strains of BGA was unsuccessful. As ostracods rapidly attain high densities soon after land preparation, they threaten the use of an algal inoculation technology. It was not established whether ostracods were responsible for the growth failure of the algal inoculant administered; for although ostracod densities were greater in inoculated plots these may have been sustained by the increased growth of indigenous flora as indicated by the number of CFU in these plots. Decomposition of inocula liberates nutrients which are probably utilised by indigenous flora. Survival of an inoculum is dependent upon many environmental factors and as such, successful introduction of non-native strains of BGA is troublesome if not impractical. It may be easier to encourage growth of indigenous N_2-fixing BGA which, by all accounts, are widespread geographically and present in many soil types.

The practices of algal inoculation and enhancement of indigenous BGA, are both in the experiment stage of development. Although many inoculation trials have been reported, none have shown conclusively that the algae inoculated was responsible for either the ensuing bloom (if present) or measured yield increases (if any). According to Venkataraman (1972) the cost of producing inoculum in plots is low but the cost of nurturing the inoculum once in the field will be additional if soil amendments and grazer control is necessary. Encouraging indigenous strains to grow or increase their biomass in the field dispenses with inoculum production. Still, the greatest obstacles to both techniques are not knowing what factors influence bloom formation and what makes a certain inoculum competitive in a foreign environment. Nevertheless, BGA already constitute a N source in neutral and alkaline soils and efforts to increase the range of soil types in which BGA activity can be intensified should be pursued.

SUMMARY

An introduction to the soil/floodwater ecosystem of lowland rice fields is given. Two primary consumers are particularly important in limiting the growth and N$_2$-fixing activities of blue-green algae in irrigated rice; the Ostracoda (Class Crustacea) and the Pulmonata (Mollusca). Control of grazing by neem seeds *Azadirachta indica* A. Juss and cultural practices enhanced BGA biomass and increased N$_2$-fixation ten fold. Significant increases in rice grain protein occur if heterocystous algae bloom early in the rice cultivation cycle and grazing control is maintained over 40 days. A large positive N balance was obtained over 3 rice crops by using neem seeds to control grazing of BGA. Algal inoculants used in conjunction with grazer control failed to establish themselves, and factors other than grazing were considered responsible. Plant-derived pesticides showed great promise for sustainable agriculture.

References

Aboul-Fahl, M., Taha Eid, M., Hamissa, M.R., El-Nawawy, A.S. & Shoukry, S. (1967). The effect of the nitrogen-fixing blue-green alga *Tolypothrix tenuis* on the yield of paddy. *Journal of Microbiology, U.A.R.*, **2**, 241–249.

Alimagno, B.V. & Yoshida, T. (1975). Growth and yield of rice in Maahas soil inoculated with nitrogen-fixing blue-green algae. *Philippines Agriculture*, **59**, 80–90.

All India Coordinated Project on Algae (1979). Algal biofertilizers for rice. Indian Agricultural Research Institute; New Delhi. 61 pp.

Amma, P.A., Aiyer, R.S. & Subramoney, N. (1966). Occurrence of blue-green algae in acid soils in Kerala. *Agricultural Research Journal of Kerala*, **14**, 141.

Bains, S.N., Prasad, R. & Bhatta, P.C. (1971). Use of indigenous materials to enhance the efficiency of fertiliser N for rice. *Fertiliser News*, **16**, 30–32.

Chen, P.C. (1983). Characterization of a dinitrogen fixing cyanobacterium isolated from a rice paddy field. *Proceedings of the National Science Council, B*, ROC.

De, P.K. (1936). The problem of nitrogen supply of rice. I. Fixation of nitrogen in the rice soils under water-logged conditions. *Indian Journal of Agricultural Science*, **6**, 1237–1245.

De, P.K. & Mandal, L.K. (1958). Fixation of nitrogen by algae in rice soil. *Soil Science*, **81**, 453–458.

De Batta, S.K., Fillery, I.R.P. & Craswell, E.T. (1938). Results from recent studies on nitrogen fertilizer efficiency in wetland rice. *Outlook on Agriculture*, **12**, 125–134.

Fenchel, T. & Blackburn, T.H. (1979). *Bacteria and Mineral Cycling*. Academic Press; London.

Gardner, W.S., Nalepa, T.F., Quigley, M.A. & Malczyk, J.M. (1981). Release of phosphorus by certain benthic invertebrates. *Canadian Journal of Fisheries & Agricultural Science*, **38**, 978–981.

Gardner, W.S., Nalepa, T.F., Slavens, D.R. & Laird, G.A. (1983). Patterns and rates of nitrogen release by benthic Chironomidae and Oligochaeta. *Canadian Journal of Fisheries & Agricultural Science*, **40**, 259–266.

Grant, I.F. & Alexander, M. (1981). Grazing of blue-green algae (Cyanobacteria) in flooded soils by *Cypris* sp. (Ostracoda). *Soil Science Society of America Journal*, **45**, 773–777.

Grant, I.F., Egen, E. & Alexander, M. (1983a). Measurement of rates of grazing of the ostracod *Cyprinotus carolinensis* on blue-green algae. *Hydrobiologia*, **106**, 199–208.

Grant, I.F., Tirol, A.C., Aziz, T. & Watanabe, I. (1983b). Regulation of invertebrate grazers as a means to enhance biomass and nitrogen fixation by Cyanophyceae in wetland rice fields. *Soil Science Society of American Journal*, **47**, 669–675.

Grant, I.F., Seegers, R. & Watanabe, I. (1984). Increasing biological nitrogen fixation in flooded rice using neem. Second International Neem Conference, Rauisch Holzhausen, F.R.G. (Proceedings in press).

Grant, I.F. & Seegers, R. (1985). Tubificid role in soil mineralisation and recovery of algal nitrogen by lowland rice. *Soil Biology & Biochemistry*. (In press).

Heckman, C.W. (1979). *Rice Field Ecology in Northeastern Thailand*. W. Junk; The Hague.

Hirano, R., Shiraishi, K. & Nakano, K. (1955). Studies on the blue-green algae in lowland paddy soil. Part 1. On some conditions for growth of blue-green algae in paddy soil and its effects on growth of the paddy rice plant. (In Japanese). *Shikoku Shijenko Hokoku*, **2**, 121–137.

Huang, C.Y. (1978). Effects of nitrogen fixing activity of blue-green algae on the yield of rice plants. *Botanical Bulletin of Academia Sinica*, **19**, 41–52.

Ichimura, S. (1954). Ecological studies on the plankton in paddy fields. I. Seasonal fluctuations in the standing crop and productivity of plankton. *Japanese Journal of Botany*, **14**, 269–279.

Kikuchi, E., Furusaka, C. & Kurihara, Y. (1975). Surveys of the fauna and flora in the water and soil of paddy fields. *Reports of the Institute of Agricultural Research, Tohoku University*, **26**, 25–35.

Kikuchi, E. & Kurihara, Y. (1982). The effects of the oligochaete *Branchuria sowerbyi* Beddard (Tubificidae) on the biological and chemical characteristics of overlying water and soil in a submerged ricefield soil system. *Hydrobiologia*, **97**, 203–208.

Konishi, C. & Seino, K. (1961). Studies of the maintenance-mechanism of paddy soil fertility in nature. *Bulletin of the Hokuriku Agricultural Experimental Station*, **2**, 41–136.

Koyama, T. & App, A. (1979). Nitrogen balance in flooded rice soils. In *Nitrogen and Rice*. pp. 95–104. International Rice Research Institute; Los Banos, Philippines.

Kurasawa, H. (1956). The weekly succession in the standing crop of plankton and zoobenthos in the paddy field (1). (In Japanese). *Shigen Kagaku Kenshyusho Iho*, **45**, 86–99.

McCullough, F.S., Gayral, Ph., Duncan, J. & Christie, J.D. (1980). Molluscicides in schistosomiasis control. *Bulletin of the World Health Organisation*, **58**, 681–689.

Okuda, A. (1948). Studies on the effect of drainage on rice crop. (In Japanese). *Agronomy*, **2**, 306.

Osa-Afiana, L.O. & Alexander, M. (1981). Factors affecting predation by a microcrustacean (*Cypris* sp.) on nitrogen fixing blue-green algae. *Soil Biology & Biochemistry*, **13**, 27–32.

Pantastico, J.B. & Suayan, A.Z. (1973). Algal succession in the rice fields of College and Bay, Laguna. *Philippines Agriculture*, **57**, 313–326.

Reynaud, P.A. & Roger, P.A. (1978). N_2-fixing algal biomass in Senegal rice fields. *Ecological Bulletin (Stockholm)*, **26**, 148–157.

Roger, P.A. & Kulasooriya, S.A. (1980). *Blue-Green Algae and Rice*. The International Rice Research Institute; Los Banos, Philippines.

Roger, P.A., Kulasooriya, S.A., Barraquio, W.L. & Watanabe, I. (1981). Epiphytic nitrogen fixation on lowland rice plants. In *Nitrogen Cycling in South East Asian Wet Monsoon Ecosystem* (R. Wetselaar *et al.*, eds.) Australian Academy of Science; Canberra.

Saito, M. & Watanabe, I. (1978). Organic matter production in rice field flood water. *Soil Science & Plant Nutrition*, **24**, 427–440.

Schmutterer, H. & Zebitz, C.P.W. (1984). Standardised methanolic extracts from seeds of single neem trees of African and Asian origin: The effect on Coleoptera and Diptera. Second International Neem Conference, Rauisch Holzhausen, F.R.G. (Proceedings in press).

Shiga, H. & Ventura, W. (1976). Nitrogen supplying ability of paddy soils under field conditions in the Philippines. *Soil Science & Plant Nutrition*, **22**, 387–399.

Singh, R.N. (1961). *Role of Blue-Green Algae in Nitrogen Economy of Indian Agriculture*. Indian Council of Agricultural Research; New Delhi. 175 pp.

Subramanyan, R.G., Manna, B. & Patnaik, S. (1965). Preliminary observations on the inter-actions of different rice types to inoculation of blue-green algae in relation to rice culture. *Proceedings of the Indian Academy of Science, Section B*, **62**, 171–175.

Venkataraman, G.S. (1972). *Algal Biofertilizers and Rice Cultivation*. Today and Tomorrow's Printers and Publishers; Faridabad, Hariyana.

Wada, H., Inubishi, K. & Takai, Y. (1982). Easily decomposable organic matter in paddy soils. III. Relationship between chlorophyll-type compounds and mineralizable nitrogen. *Japanese Journal of Soil Science & Plant Nutrition*, **53**, 380–384.

Watanabe, A., Nishigaki, S. & Konishi, C. (1951). Effect of nirogen-fixing blue-green algae on the growth of rice plant. *Nature (London)*, **168**, 748–749.

Watanabe, A., Ito, R. & Sasa, T. (1955). Micro-algae as a source of nutrients for daphnids. *Journal of General & Applied Microbiology*, **1**, 137–141.

Watanabe, A. (1973). On the inoculation of paddy fields in the Pacific area with nitrogen-fixing blue-green algae. *Soil Biology & Biochemistry*, **5**, 161–162.

Watanabe, I., Lee, K.K. & de Guzman, M. (1978). Seasonal Change of N$_2$ fixing rate in rice field assayed by *in situ* acetylene reduction technique. II. Estimate of nitrogen fixation associated with rice plants. *Soil Science & Plant Nutrition*, **24**, 465–471.

Watanabe, I. & Roger, P.A. (1984). Nitrogen fixation in wetland rice field. In *Current Developments in Biological Nitrogen Fixation* (N.S. Subba Rao, ed.) pp. 237–276. Oxford and I.B.H. Publishing Co.; Delhi.

Webbe, G. & Lambert, J.D.H. (1983). Plants that kill snails and prospects for disease control. *Nature (London)*, **302**, 754.

Wilson, J.T., Greene, S. & Alexander, M. (1980). Effect of micro-crustaceans on blue-green algae in flooded soil. *Soil Biology & Biochemistry*, **12**, 237–240.

Yamaguchi, M. (1979). Biological nitrogen fixation in flooded rice field. In *Nitrogen and Rice*. Pp. 193–203. International Rice Research Institute; Los Banos, Philippines.

The Search for Biological Control Agents against Plant Pathogens: A Pragmatic Approach

R. Campbell

Department of Botany, University of Bristol, Bristol, BS8 1UG, U.K.

INTRODUCTION

This paper will not present any details of research or review previous work, but will try to set out some criteria for the use of biological control agents in a sustainable agriculture. After many years of work there is now a considerable, though still inadequate, theoretical background to the biological control of plant pathogens and a large number of examples of the experimental use of biological control agents (Baker & Cook, 1974; Cook & Baker, 1983). There are however very few examples of the control of plant pathogens by other, antagonistic organisms that are either available or used in commercial agriculture (Cook & Baker, 1983). It is perhaps time to consider briefly why this may be so and under what conditions biological control, by means of introduced antagonists, might successfully be used. How might such organisms be selected? This may help those who are considering whether some form of biological control may be of use in a particular situation, and may cause those of us actively engaged in developing biocontrol agents to stand back from the details and see if the aims and the approach are logical. The economic and environmental pressures for the development of biological agents have changed since a previous examination of these problems (Baker & Cook, 1974).

It is unlikely that introduced biocontrol agents will, in the near future anyway, be the sole agents used to control plant diseases. Rotation of crops, and the many tillage systems that give various degrees of protection or benefit to the soil, are important to ensure that the plants grow well and are adequately supplied with nutrients, which may reduce the deleterious effects of some diseases without actually affecting the pathogen. In the agricultural systems of the 'developed' countries, the biocontrol agents are likely to be used alongside pesticides and other agrochemicals in various forms of integrated control.

Chemicals can be designed or selected to kill fungi and to control diseases and, initially at least, they may give spectacularly good results with complete removal of the unwanted pathogen. Such pesticides are the result of an enormous investment in time, resources and money by the industrialized nations and international companies. Large chemical companies may screen 8,000 or 10,000 possible chemicals a year in order to find one marketable product (Delp, 1977; Finch, 1979) and many of them have been doing this for perhaps 40 years.

In contrast biocontrol agents rarely eliminate disease, they reduce symptoms to acceptable levels and maintain a balance between the pathogen and the antagonists in this situation. The pathogen is usually present, indeed it may be essential to the long term survival of the control agent. Biological control therefore tends not to produce such rapid, dramatic, answers to disease problems compared with an effective chemical control. One of the main reasons why biocontrol systems have not been developed in competition with chemicals is that, in comparison, they have had insignificant amounts of money and other resources devoted to their development. Until recently biocontrol agents were typically developed, almost as a spare time occupation, from incidental observations or anecdotal evidence (e.g. Fox, 1965). There has been very limited systematic search for possible agents, with usually only a few hundred organisms being looked at before the selection is made. There has been a recent increase in the interest in biological control, especially the use of introduced antagonists, but in general investigations are still on a small scale.

There is no obvious reason why the systematic, extensive search for biocontrol agents, and the subsequent testing and formulation of a potential product, will be any easier or cheaper than the development of the present pesticides has been. Indeed the present attempts to do biological control 'on the cheap' may have been one of its major problems.

WHEN AND WHERE CAN BIOLOGICAL AGENTS BE USED?

There may be long term aims for the production of biocontrol agents as a more 'natural' and non-polluting system for the control of plant pathogens. This may come about because of political pressure from various groups to restrict the use of pesticides or even to ban them completely.

There are also technical reasons why biological control may in the future become more important. A great many chemicals have been screened to produce the presently available fungicides, but pathogens develop resistance to them (Georgopoulos, 1977). There is a finite number of compounds and it may be that most of the potentially useful chemical groupings have been found and developed. The search for new products is likely to be ever more difficult, to take longer and therefore to cost more. There is a considerable amount to be done by changing formulations or making minor chemical changes to molecules which are known to effective pesticides but this is only delaying the day when fungicides will be less effective than they are now. Other methods of control, including biological ones, will then be needed, rather than just being desirable.

It is not however likely that such considerations will finance or initiate biological control programmes now. There are more pragmatic aims that must be fulfilled by biocontrol agents if they are to meet the present situation in agricultural systems which are dominated by chemical controls of pathogens. Will the proposed biological agent control diseases that chemicals or plant breeding cannot control, or will it do it more effectively or more cheaply? For example there are many hundreds of possible biological control agents reported in the academic literature for the control of foliar diseases of agricultural crops (Blakeman & Fokkema, 1982) but none are used commercially. Since leaves are accessible there are many effective, relatively cheap and easy to use, foliar fungicides. In contrast there is a lack of chemical control agents for root diseases of field crops, partly because pesticides are difficult to apply to below-ground organs of the plant. Root diseases could be a rewarding area for research.

Biocontrol agents that are more trouble or more expensive just will not be employed, unless legislation enforces their use or prohibits pesticides for that disease or situation. Those biological agents that are too complex or which demand advanced technology will not be available in many underdeveloped countries.

There are other criteria which might suggest candidates for the possible immediate use of biocontrol agents. It is relatively easy to introduce foreign organisms into a vacant ecological niche, but competition may make it very difficult to establish them in an existing, occupied, stable ecosystem such as soil. Thus nursery composts are almost sterile and can be inoculated with antagonists with a reasonable chance of a successful establishment: for example *Trichoderma* is used to control damping-off (by *Pythium* and *Rhizoctonia*) and *Sclerotinia* rot of onions (Corke & Risbeth, 1981; Henis, 1984). There is another advantage to using horticultural crops in that they are often grown in relatively controlled environments such as glasshouses and it may be possible to modify the conditions to suit the antagonist or predator rather than the pathogen. Furthermore horticultural crops are often intensively grown and of high value so they can use rather expensive or complex control measures.

Another vacant niche situation occurs in tree trunks which have a very low level of microbial colonization when healthy. *Trichoderma* has been used to control *Chondrostereum purpureum* in fruit trees (Corke, 1974). Similarly *Peniophora (Phlebia) gigantea* controls *Heterobasidion annosum* when applied to the freshly cut surfaces of some conifer tree stumps (Corke & Risbeth, 1981). *Peniophora* does not work if the stump has already been colonized by the pathogen or by other saprophytes, it needs a vacant niche.

A final consideration in selecting suitable diseases for biological control is to obtain any anecdotal evidence or incidental observations which may suggest that biological control is already operating. This may be an observation that when the disease is not serious there is often a particular organism or association of organisms present. The use of fungicides which kill saprophytes and are not very effective against the pathogen may provide evidence: the disease may get worse after a spray (Blakeman & Fokkema, 1982; Griffiths, 1981) if the removal of the competing saprophytes reduces disease control more than the toxic effect of the fungicide reduces the pathogen.

It may therefore be possible to select diseases that are particularly suitable for biological control by the application of antagonists because the disease itself is important and not at present effectively controlled, or for ecological, political or economic reasons. If there are scientific reasons to think that it may be easy or even possible to successfully introduce biocontrol agents into the system, then all that remains is to select the organism in a logical and effective way! What criteria should be used to do this?

CRITERIA AND METHODS FOR SELECTION AND TESTING OF BIOCONTROL AGENTS

The generalized system is shown in Figure 1, though the details will need to be adapted for each particular disease and proposed control system. The aim is to test the biocontrol agent in stages, with the most severe tests at the start to eliminate unsuitable organisms as soon as possible. The costs in the system increase greatly as the organisms pass through the system. Thus it is relatively cheap to isolate and screen the potential organisms, rather more expensive to investigate them and very expensive indeed to run full scale field trials and to do full toxicity testing and patenting. It is assumed that there will be a useful, commercial product at the end of the sequence and the tests are designed accordingly. If organisms are wanted for academic exercises then the criteria of selection will be different.

Search Sites

The selection of search sites for the organisms (Fig. 1) is of greatest

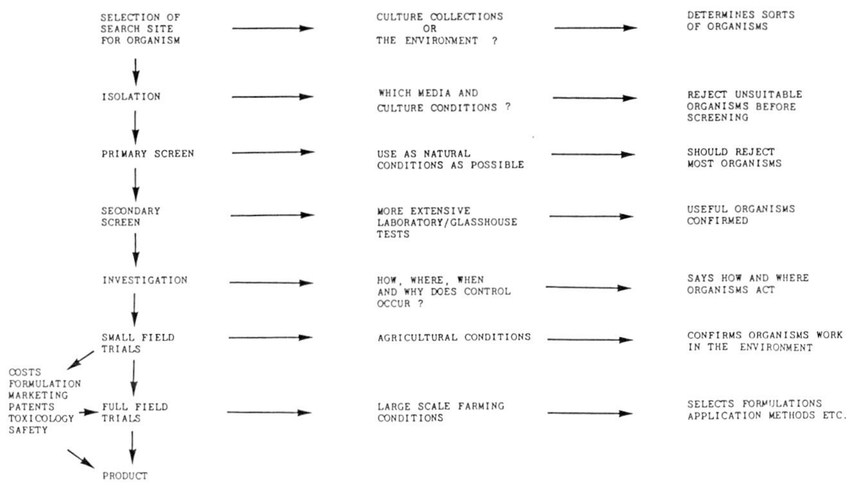

FIGURE 1   The steps necessary to select and test potential biocontrol agents.

importance for it determines what will be found. In general, culture collections are not suitable: they will contain organisms adapted to high nutrient conditions and unlikely to survive in nature. Collections may be useful if a particular group of organisms is of interest: for example it may be worth checking all isolates of a genus of proven ability.

The best source of isolates is healthy plants, or at least those that show few symptoms, in conditions or situations where the pathogen is present and the disease would normally be expected to occur. There could be something preventing infection or symptom expression in these conditions. It is not good policy to look for antagonists on heavily diseased plants or crops, for clearly there is nothing in the environment that is preventing the pathogen from affecting the plant. Biocontrol agents against root diseases may be looked for in suppressive soils, either naturally occurring ones or those induced by repeated cropping with the same plant. Places where diseases occur in patches and the spread has been halted, or apparently healthy plants in otherwise heavily diseased crops, may be useful. Similarly wounds to trees that have healed well, rather than rotting, may have been colonized by organisms antagonistic to those causing decay. Infections by mycoparasites may occur on the pathogen and this may reduce disease or disease spread and they are obviously useful organisms. Again a slightly diseased plant is necessary for the control agent to operate so such biocontrol will never give zero disease.

Search sites should be subject to the conditions that the potential control agent will have to operate under. They should be in a reasonably similar climatic zone on the same or similar host plants, and subject to normal agricultural practice. The latter may well include the use of pesticides: it is worthless to produce a good control agent that is destroyed by the first pesticide spray applied to that or neighbouring crops.

Isolation

The isolation of the potential biocontrol agents is not usually difficult microbiologically. In general dilute, simple media should be used so that any organisms can later be grown easily and cheaply in industrial fermenters. It may be decided to select spore-forming organisms by pasteurization of the initial substrate: this will give good shelf life in the final product if that is considered important. Selecting and developing even the most efficient antagonist is useless if the organism is so fastidious and has such exotic requirements that it can never be produced economically or survive commerical processes. It may be advantageous to isolate particular groups, such as fluorescent pseudomonads, by selecting them on specialized media, if they have shown themselves to be useful in the past.

The above methods only apply to saprophytes. Biotrophs, such as mycorrhizal fungi, may also give protection from pathogens (Harley & Smith, 1983). Ectomycorrhizas can be cultured and could therefore use the methods just described with specialized media, but the important vesicular arbuscular mycorrhizal fungi cannot be grown in culture. It may be possible to isolate on live bait plants but this severely restricts the numbers that could be handled. The isolation and the primary screen may be combined by inoculating live plants with root macerations for biotrophs, and also inoculating the disease organism, then examining those plants which are colonized by biotrophs and not killed by the pathogen. Such systems would be difficult to use on a large scale.

Primary screen

This should be easy to do, cheap and rigorous. By analogy with screens for the selection of pesticides, which have undoubtly been successful in selecting chemicals to kill organisms, the biological primary screen may have to find 1 organism in 1000 or even 10,000. It must fail at least 99.9% of organisms and so must handle a large number to have a chance of success. Little replication is used; the effect of the best organisms should be so obvious as not to require statistical proof. Screens based on *in vitro* growth in petri dishes are best avoided if possible. They are likely to select organisms which work well in culture, by antibiotic production for example, but which may not produce such substances in the low nutrient conditions of the natural environment.

If at all possible live plants growing in the presence of the pathogen should be the basis of the primary screen, and no particular effect should be studied. If the plant grows better or is healthier then the organism is potentially useful, no matter whether it works by antibiotic production, lysis of the pathogen, production of plant growth promoting substances or anything else. It may also be worth putting the potential biocontrol agents onto healthy plants to look for growth promoting organisms in the absence of disease. In soil diseases the test may be on whole seedlings in small pots (to reduce the space requirements with large numbers). Individual leaves can be used for leaf diseases so that many tests can be done on one larger plant.

Secondary screen

The secondary screen (Fig. 1) merely looks a little more closely at the effects shown by the primary screen. There may be more replicates for statistical tests, with the plants growing in more natural conditions and to a larger size. Many fewer potentially useful organisms are involved at this stage so more elaborate assays are also possible, perhaps with a quantitative assessment of the disease rather than just looking for plants that are obviously better as was done in the primary screen.

Any organisms which pass this screen are potentially useful and worthy of further investigation.

Investigation

It is now necessary to know a little more about the organisms which have passed through the initial stages of selection. They may be identified, if possible, so that effective literature searches can be made. It is also essential to show that the agents are not plant pathogens themselves, and are not animal or human pathogens that might be a danger, especially as large scale inoculum production might now be needed.

The mode of action must also be demonstrated and the stage in the life cycle of the pathogen which is attacked must be identified. This will enable a logical development of delivery systems. Does the organism need to be on the leaf before the pathogen arrives? Does it spread on the plant or has the initial application got to achieve complete cover? Does the biocontrol agent persist in the environment? There may be organisms such as protozoa that eat the control agent or resident flora that antagonize it. What density or concentration of inoculum is needed? Does the antagonist have to contact the pathogen or can it operate from a distance? It is no good putting an organism on the crop when the disease develops if that organism acts by lysing resting structures. Similarly it would be needlessly expensive to try to achieve complete cover of a plant if the organism is itself mobile enough to move 10 cm.

The answers to these questions, and indeed the questions themselves, will obviously vary with the particular disease, but it is likely that detailed investigations with the light microscope, and possibly the electron microscope if the biocontrol agent is a bacterium, will be needed. Gnotobiotic plants (grown from surface sterilized seed, with known organisms added) may be used to show exactly how the antagonist operates and to prove its efficacy. It is possible to hypothesise that the potential control agent is in fact doing nothing other than provide a nutrient source for some organism in the environment and it is this second organism which controls the disease. Only by having a sterile plant and then adding the pathogen and the antagonists can the action, direct or otherwise, be proved. The study may also require the use of control agents with antibiotic resistance so that they can be isolated from the soil population, by using selective media, for the study of persistence or colonization ability. The production of strain specific antibodies and subsequent fluorescent labelling of the introduced strain is another technique for following the introduced organisms in the natural soil population.

The investigation may take a long time and testing in small field trials can continue on an empirical basis meanwhile, though refined and corrected as the results of the investigations become known.

Small scale field trials

These trials (Fig. 1) are the first test under anything near to normal conditions. They are still quite expensive and may be done with several potential antagonists, so the plot size is usually small, possibly only a few metres square for field crops or only on a part of a tree. High replication is usually needed for the variance of the data is much greater under field conditions than in most glasshouse experiments. Normal agricultural sprays, fertilizers, tillage and cultivation should be used if possible, to test the organisms under the most realistic conditions. Several sites, with different soils or exposure, are needed to give indications of problems with clay soils (by adsorbing antibiotics) or with the survival of the antagonists under unfavourable environmental conditions.

Full field trials

These can be very expensive in time, money and personnel and are only done when it reasonably certain that the organisms may be successful. What is really being tested is not the organism itself, but formulations, dose rates and effects of different agricultural practices. Combinations of organisms may be investigated in this final fine-tuning of the system. These mixed formulations may allow an extension of the working range by combining agents with different characteristics; for example bacteria with the ability to operate at low water tensions in the soil may be combined with those favoured by wetter conditions so that the formulation will work over a range of water potentials. For foliage crops the combination might be resistance to several different pesticides or the ability to control a range of diseases. Integrated control, that is the use of combinations of chemical and biological methods, must also be considered at this stage.

The extensive nature of the trials may mean that the detailed disease assessments may no longer be possible, especially with root diseases, but good data for effects on yield should be obtained.

Alongside the full field trials there will also be extensive work on the investigation of the organism, final pathogenicity and toxicity testing against humans and animals, the preparation of patents, etc. (Fig. 1). This can be very expensive, especially if the biocontrol agents are considered as agrochemicals and required to undergo the full environmental and toxicity testing.

CONCLUSIONS

The main part of the above discussion has been concerned with the selection of sites, the isolation and the screening techniques. These are the areas that have been most neglected and demand most study, for the investigative methods and the field trials are well provided with methods and traditional expertise.

Genetic engineering is now in vogue and has been applied to some agricultural problems, most obviously to nitrogen fixation (Dixon *et al.*, 1981) and the selection and development of new and better strains of *Rhizobium*. It is however premature in the present studies of biological control. The first requirement is for a good organism and this will be obtained by screening natural populations. When the biological control agent is produced, it may be useful to consider various modifications such as altering the nutrient requirements, growth characteristics, etc. to improve the fermentation system. Alternatively, or in addition, genetic engineering may be used to improve mobility, growth characteristics in the soil, pesticide or antibiotic tolerance or reduction in predation. It is not likely that genetic engineering will be used in the near future to do anything but modify an existing effective organism. There is just not enough known about the microbial ecology of the plant microbial interaction to 'design' an organism: organisms will have to be selected empirically and then improved a little.

Another possibility is that the host may be modified to suit the biocontrol agent. It is known that different plants have different microbial populations (Newman *et al.*, 1974) and so presumably favour some organisms rather than others. It may be possible to breed a plant which favours certain root surface bacteria because of its exudates: it is already known that wheat can be made to change the rhizosphere population, and increase the number of antagonistic organisms present, by standard plant breeding techniques (Neal *et al.*, 1970). Perhaps it may be possible in the future to breed or design the host and the biocontrol agent together. Soil additives may also modify the microbial flora (Cook *et al.*, 1978; Cook & Baker, 1983) and could be used in combination with the proposed biocontrol agent. In studies on leaves it is possible to change their surface characteristics by breeding or selecting new cultivars (Baker, 1974) or to alter the microclimate by affecting leaf shape or attitude, so again it may be possible to look at antagonists and host together.

There are now examples, both experimental and commercial, of biological control with introduced antagonists, but not on a large scale. The main reason for this is simply a lack of investment of time and money to develop the systems of screening, formulation and production. If a very small proportion of the resources which have been spent on pesticides had been devoted to biological control we would already have a selection of comercially available organisms. However the situation is now changing from several different directions. There is beginning to be some understanding of the role of microorganisms in and on both healthy and diseased plants. There is the political pressure which has arisen from ecological problems with pesticides and intensive fertilizer use. Thirdly there are the twin problems of over-production in the intensive, high input western agriculture, and under production in the impoverished low input agriculture of the developing world. It is clear that a new balance will eventually have to be struck, with a less intensive agriculture in the industrialised countries, and more productive, but not necessarily more costly, systems must be developed for the rest of the world. Both must be sustainable, in that they do not require massive imputs of energy and artificial, often oil-based, chemicals. Biological control clearly has a place in this development since in the ideal situation it will maintain a balance between the pathogens and the natural or introduced antagonistic organisms, and will reduce if not avoid the use of toxic chemicals for the control of plant diseases.

References

Baker, E.A. (1974). The influence of environments on leaf wax development in *Brassica oleracea* var. *gemmifera. New Phytologist*, **73**, 955–966.

Baker, K.F. & Cook, R.J. (1974). *Biological Control of Plant Pathogens*. Freeman; San Francisco. 433 pp.

Blakeman, J. & Fokkema, N.J. (1982). Potential for biological control of plant diseases on the phylloplane. *Annual Review of Phytopathology*, **20**, 167–192.

Cook, R.J., Boosalis, M.G. & Doupnik, B. (1978). Influence of crop residues on plant diseases. In *Crop Residue Management Systems* (W.R. Oschwald, ed.), pp. 147–163. American Society for Agronomy, Special Publication No. 31; Madison, Wisconsin.

Cook, R.J. & Baker, K.F. (1983). *The Nature and Practice of Biological Control of Plant Pathogens.* American Phytopathological Society; St. Paul, Minnesota. 539 pp.

Corke, A.T.K. (1974). The prospect for biotherapy in trees infected by silver leaf. *Journal of Horticultural Science*, **49**, 391–394.

Corke, A.T.K. & Risbeth, J. 1981. Use of microorganisms to control plant diseases. In *Microbial Control of Plant Pests and Diseases 1970–1980* (H.D. Burges, ed.), pp. 717–736. Academic Press; London.

Delp, C.J. (1977). Privately supported disease management activities. In *Plant Disease* (J.G. Horsfall & E.B. Cowling, eds.), Vol. 1, pp. 381–392. Academic Press; New York.

Dixon, R., Kennedy, C. & Merrick, M. (1981). Genetic control of nitrogen fixation. In *Genetics as a Tool in Microbiology* (S.W. Glover & D.A. Hopwood, eds.), pp. 161–185. Cambridge University Press; Cambridge.

Finch, R., ed. (1979). *Pesticides–the Chemical Control of Crop Pests and Diseases.* Imperial Chemical Industries Ltd. 17 pp.

Fox, R.A. (1965). The role of biological eradication in root disease control in replantings of *Hevea brasiliensis*. In *Ecology of Soil-Borne Plant Pathogens* (K.F. Baker & W.C. Snyder, eds.), pp. 348–373. Murray; London.

Georgopoulos, S.G. (1977). Pathogens become resistant to chemicals. In *Plant Disease* (J.G. Horsfall & E.B. Cowling, eds.), Vol. 1, pp. 327–345. Academic Press; New York.

Griffiths, E. (1981). Iatrogenic plant diseases. *Annual Review of Phytopathology*, **19**, 69–82.

Harley, J.L. & Smith, S.E. (1983). *Mycorrhizal Symbiosis.* Academic Press; London. 483 pp.

Henis, Y. (1984). Ecological principles of biocontrol of soilborne plant pathogens: *Trichoderma* model. In *Current Perspectives in Microbial Ecology* (M.J. Klug & C.A. Reddy, eds.), pp. 353–361. American Society for Microbiology; Washington.

Neal, J.L., Atkinson, T.G. & Larson, R.I. (1970). Changes in the rhizosphere microflora of spring wheat induced by disomic substitution of a chromosome. *Canadian Journal of Microbiology*, **16**, 153–158.

Newman, E.I., Campbell, R., Christie, P., Heap, A. & Lawley, R.A. (1979). Root microorganisms in mixtures and monocultures of grassland plants. In *The Soil-Root Interface* (J.L. Harvey & R. Scott Russell, eds.), pp. 161–173. Academic Press; London.

Index